T0280710

CAMBRIDGE LIBRARY COLLECTION

Books of enduring scholarly value

Darwin

Two hundred years after his birth and 150 years after the publication of 'On the Origin of Species', Charles Darwin and his theories are still the focus of worldwide attention. This series offers not only works by Darwin, but also the writings of his mentors in Cambridge and elsewhere, and a survey of the impassioned scientific, philosophical and theological debates sparked by his 'dangerous idea'.

Monographs on the Fossil Lepadidae, Balanidae and Verrucidae

These two short monographs, published under the auspices of the Palaeontographical Society in 1851 and 1854, show Charles Darwin as a meticulous research scientist, poring over fossils collected by himself and other enthusiasts in Britain and in Europe. The first volume is devoted to the Lepadidae, and the second to the Balanidae and Verrucidae (all types of barnacle, members of the infraclass Cirripedia). Darwin's interest in barnacles had first arisen in his student days in Edinburgh, under the guidance of Robert Grant, and increased during his detailed work in dissecting and classifying the specimens he had collected on the Beagle voyage. The publication of his findings cemented his reputation as a expert taxonomist and biologist, and his observations over eight years of the minute differences between males, females and an apparent hermaphroditic stage of development lent support to his developing theory of evolution.

Cambridge University Press has long been a pioneer in the reissuing of out-of-print titles from its own backlist, producing digital reprints of books that are still sought after by scholars and students but could not be reprinted economically using traditional technology. The Cambridge Library Collection extends this activity to a wider range of books which are still of importance to researchers and professionals, either for the source material they contain, or as landmarks in the history of their academic discipline.

Drawing from the world-renowned collections in the Cambridge University Library, and guided by the advice of experts in each subject area, Cambridge University Press is using state-of-the-art scanning machines in its own Printing House to capture the content of each book selected for inclusion. The files are processed to give a consistently clear, crisp image, and the books finished to the high quality standard for which the Press is recognised around the world. The latest print-on-demand technology ensures that the books will remain available indefinitely, and that orders for single or multiple copies can quickly be supplied.

The Cambridge Library Collection will bring back to life books of enduring scholarly value across a wide range of disciplines in the humanities and social sciences and in science and technology.

Monographs on the Fossil Lepadidae, Balanidae and Verrucidae

CHARLES DARWIN

CAMBRIDGE UNIVERSITY PRESS

Cambridge New York Melbourne Madrid Cape Town Singapore São Paolo Delhi

Published in the United States of America by Cambridge University Press, New York

www.cambridge.org
Information on this title: www.cambridge.org/9781108004824

This edition first published 1851
This digitally printed version 2009

ISBN 978-1-108-00482-4

THE

PALÆONTOGRAPHICAL SOCIETY.

INSTITUTED MDCCCXLVII.

MDCCCLI.

A MONOGRAPH

ON THE

FOSSIL LEPADIDÆ,

OR,

PEDUNCULATED CIRRIPEDES OF GREAT BRITAIN.

BY

CHARLES DARWIN, F.R.S., F.G.S.

LONDON:

PRINTED FOR THE PALÆONTOGRAPHICAL SOCIETY.

1851.

PREFACE.

I HAVE great pleasure in returning my most sincere thanks to various naturalists, both for intrusting to me their collections of Fossil Cirripedia, and for allowing me, whenever it was advisable, to clear the specimens from their matrix. Although an entire stranger to many of the gentlemen to whom I applied, I have in every instance received the most courteous acquiescence to my demands. To Mr. Fitch, of Norwich, I here beg to return my thanks, for having allowed me to keep, during several months, his unrivalled collection of Cirripedia from the Upper Chalk of Norwich,—the fruit of twenty years' labour. Mr. Bowerbank has given me the freest use of his fine collection, rich in specimens from the Gault. Mr. Wetherell placed in my hands his beautiful and unique specimen of *Loricula pulchella*, and other species. Professor Buckman sent me, of his own accord, a fine series of the valves of *Pollicipes ooliticus*, the most ancient Cirripede as yet known, discovered and named by him. To Messrs. Flower, Searles Wood, F. Edwards, Harris, S. Woodward, Tennant, and other gentlemen, I owe the examination of several species new to me. Mr. Morris and Professor E. Forbes have, in their usual kind manner, supplied me with much valuable information, and with the loan of many specimens. To Mr. James de C. Sowerby I must express my thanks for the valuable aid rendered to me by the loan of the original specimens figured in the 'Mineral Conchology;' and for the pains exhibited in the drawings here published.

Professor Forchhammer, of Copenhagen, not only placed at my disposal many valuable specimens deposited in the Geological Museum of the University, but applied to Professor Steenstrup, who, in the most generous manner, sent me the collection in the Zoological department, including the highly valuable original specimens of his excellent Memoir on the Fossil Cirripedia of Denmark and Scania. Subsequently, Professor Steenstrup sent me a second large collection, the fruit of the indefatigable labours of M. Angelin, in

Scania: all these northern specimens have been of the greatest use to me in illustrating the British species. Having applied to Professor W. Dunker, of Cassel, for some of the species described by various German authors, he not only sent me many specimens out of his own collection, but procured from Messrs. Roemer, Koch, and Philippi, other specimens of great value; and to these most distinguished naturalists I beg to return my very sincere thanks. Lastly, I may be permitted to state, that I hope very soon to have another and more appropriate opportunity of publicly expressing my gratitude to various gentlemen, who for many months together have left in my hands their large and valuable collections of recent Cirripedia, and who have assisted me in every possible way. I will here only state, that it was owing to the suggestion and encouragement of Mr. J. E. Gray, of the British Museum, that I was first induced to take up the systematic description of the Cirripedia, having originally intended only to study their anatomy. To all the foregoing gentlemen, I shall ever feel under the deepest obligations.

INTRODUCTION.

THE CIRRIPEDIA, both recent and fossil, have been much neglected by systematic naturalists: the fossil species have, however, been more attended to than the recent. Professor Steenstrup has published[1] an excellent monograph on the Danish and Scanian Cretaceous species: Mr. J. de Carle Sowerby has given good plates of several British valves in the Mineral Conchology; and F. Roemer[2] has illustrated, by rather indifferent figures, though clear descriptions, various German forms. Other less important notices have appeared by several authors. As yet, however, no monograph has been produced on the whole group. The present volume is confined to the *Lepadidæ* or Pedunculated Cirripedia; and it so happens that the introduction, under the form of notes, of a few foreign species (which are necessary to illustrate the British species), serves to render this Monograph tolerably complete; that is, as far as the specimens collected on the Continent (judging from published accounts) serve for this end,—for we shall immediately see that certain valves are requisite in each genus.

It is unfortunate how rarely all the valves of the same species have been found coembedded; it is evident that, with the exception of some few species, the membrane which held the valves together, decayed very easily, as it does in recent Pedunculated Cirripedes. Hence, in the great majority of cases, the several valves have been found separate. Hitherto it has been the practice of naturalists to attach specific names indifferently to all the valves; and as in each species there are from three to five or six different kinds of valve, there would have been, had not the whole group been much neglected, so many names attached to each species. On the other hand, it has occurred in several instances, that many valves belonging to quite different species have been grouped together under the same name. To avoid these great evils, I have fixed on the most characteristic valves, one in each of the two main genera, and taking them as

[1] Naturhistorisk Tidsskrift, af H. Kröyer, 1837 and 1839.

[2] Die Versteinerungen des Norddeutschen Kreidegebirges, 1841.

typical, have never, except in one instance where several valves were known all to belong to the same individual, and in another instance in which a valve was very remarkable, attached a specific name to any other one. I have, however, in two cases retained names already given to certain other valves, as they presented remarkable characters, and were almost certainly distinct. In Scalpellum I have taken the Carina or Keel-valve (*i. e.* dorsal valve of most authors) as typical; and in Pollicipes, the Scuta (*i. e.* the inferior lateral valves of most authors): it would have been desirable to have taken the same valve in both genera; but it so happened that the Carina has been much more frequently collected than any other valve in Scalpellum, in which genus it is highly characteristic; whereas in Pollicipes, it is apt to present less striking characters than the Scuta, which are, moreover, commoner in most collections. In almost all the Lepadidæ the Terga (*i. e.* the upper or posterior lateral valves) are not characteristic, and are particularly liable to variation. Although only certain valves in each genus thus receive specific names, yet from the conditions of embedment, several of the other valves can often be safely attributed to the same species.

Much confusion in nomenclature will, I think, be avoided by the plan here adopted; but the study of Fossil Cirripedia must, I fear, owing to the variability of the valves, as seen in some fossil species, and as inferred from what so commonly occurs with recent species, ever remain difficult. In very many of those recent species, of which large series have passed through my hands, several of the valves have varied so much, that had I seen only certain specimens from the opposite poles of the series, I should unhesitatingly have ranked them as quite distinct species: on the other hand there are some recent forms—for instance, some species of *Lepas*, and again *Pollicipes cornucopia*, and *elegans* of Lesson—which are perfectly distinct, but which it would be hopeless to attempt discriminating when fossilized, without quite perfect specimens. It should be borne in mind, that the recognition of the Fossil Pedunculated Cirripedes by the whole of their valves and peduncle, is identical with recognising a Crustacean by its carapace, without the organs of sense, the mouth, the legs, or abdomen: to name a Cirripede by a single valve is equivalent to doing this in a Crustacean by a single definite portion of the carapace, without the great advantage of its having received the impress of the viscera of the included animal's body: knowing this, and yet often having the power to identify with ease and certainty a Cirripede by one of its valves, or even by a fragment of a valve, adds one more to the many known proofs of the exhaustless fertility of Nature in the production of diversified yet constant forms.

I must allude to one more unfortunate cause of doubt in the classification of the extinct Lepadidæ, namely, the difficulty in attributing the separated valves to the two main genera of Scalpellum and Pollicipes; for the chief distinction between these two close genera in the recent state, lies in the number of the valves, and this can very rarely be ascertained in fossil specimens. At first I determined to follow those authors who have united both genera under Pollicipes; but reflecting that I had twelve recent and above thirty-seven fossil species, with almost the certainty—as we shall presently see—of very many more being discovered, this plan seemed to me too inconvenient to be followed. There are six recent species which I intend, in a future work, to include under Scalpellum. Four of them have been raised by Dr. Leach and Mr. Gray to the rank of genera; two other unnamed species have certainly equal, if not stronger, claims to the same rank; so again the six recent species of Pollicipes have similar claims to be divided into three genera, thus making nine genera for the twelve recent species of Scalpellum and Pollicipes. In the majority of cases it would be eminently difficult to allocate the fossil species in these nine genera; nevertheless, taking the characters necessarily used for the generic divisions of all the other recent Pedunculated Cirripedes, there can be no doubt that the formation of the above nine genera might be justified, that is, if we are allowed to neglect mere classificatory utility as an element in the decision, and further, if we are invariably bound to make as far as possible all genera of exactly the same value. As far as utility in classification is concerned, it appears to me clear that the institution of so many genera, until many more species are discovered, is highly disadvantageous: with respect to making all genera of *exactly* equal value, this, though eminently desirable, appears to me almost hopeless; I know not how to weigh the value of slight differences in different valves; or whether a difference in the maxillæ or mandibles be the more important: anyhow, in this particular case, if we raised the six recent species of Scalpellum into six genera, they assuredly would not be distinct to an exactly equal degree. Under these circumstances I have followed a middle term, and kept Scalpellum and Pollicipes distinct,—genera easy to be recognised in a recent state,—which renders the classification of the fossil species, though always difficult and liable to many errors, somewhat easier than if both genera were united into one, and much easier than if the above nine genera were admitted.

Aptychus.

Before passing to more general considerations, I must offer a few remarks on the genus *Aptychus*, or *Trigonellites*, inasmuch as quite lately a distinguished naturalist, M. D'Orbigny,[1] has adopted, and with much ingenuity supported, the view that these anomalous bodies are Pedunculated Cirripedia. It cannot be denied that the general form and lines of growth closely resemble those of the Scuta or lateral inferior valves in Lepas or Anatifa: nor can it be denied, from what we know of recent species, that the Terga (upper lateral valves) and Carina (dorsal valve), which on M. D'Orbigny's view must be considered as absent, are the most likely valves to disappear from abortion. But there are points of difference which, as it appears to me, are of far greater importance than the

[1] Cours Élémentaire de Paléontologie, 1849, vol. i, p. 254.

resemblance in mere outline. The peculiar cancellated structure, which is almost visible on the external surface even to the naked eye, is wholly unlike anything known amongst Cirripedia; a thin polished slice of the valves of Lepas and of Aptychus, viewed under a high power, are as unlike as anything can well be.[1] In Aptychus the lines of growth are conspicuous on the inner or concave surface, and indistinguishable or not plain on the outer surface; whereas in Lepas exactly the reverse holds good. Again, in some specimens it appears, that additions are made to the shell on the exterior edge of the growing margin, instead of on the inner edge, as in Cirripedia. In *Aptychus latus*, there is a rather deep internal fold along the whole of that margin, through which the cirri are supposed to have been protruded, and this is unlike anything which I have met with in Cirripedes. In all the species of Aptychus, the two valves are much the most frequently, though not invariably, found widely opened, and attached together, either exactly or nearly so, by the two margins through which the cirri must have been protruded. Now in all true fossil pedunculated Cirripedes, the valves are found either separate, which is the commonest case, or when held together, those on the opposite sides almost exactly cover each other, for there is nothing in the structure of Cirripedia tending to open the valves like the ligament in bivalve shells. How comes it, then, that the specimens of Aptychus, even those found within the protected chambers of Ammonites, thus generally have their valves widely gaping? Even if we pass over this difficulty, is it not strange that the valves should always have been held together by that margin, which in the recent condition is supposed to have been open for a considerable portion of its length, for the exsertion of the cirri; whereas, in not one single instance, as far as I have seen, are the two valves held together by the opposite margin, which in the recent state, on the idea of Aptychus having been a Cirripede, must have been continuously united by membrane.

There is another argument against Aptychus having been a Cirripede, which will have weight, perhaps, with only a few persons: in Pollicipes, the main growth of all the valves is downwards; in Lepas or Anatifa, as well as in most of the allied genera, the main growth of the Scuta and of the Carina (*i. e.* lower lateral, and dorsal, or valves,) is in a directly reversed direction, or upwards. Now Pollicipes is the oldest known genus of Cirripedes, having been found in the Lower Oolite, whereas hitherto Lepas is not certainly known to have been discovered even in the newest Tertiary formation. So again within the limits of the genus Scalpellum, I know of only two cretaceous species in which the Scuta grow upwards and downwards, and only one case in which the Carina has this double direction of growth; whereas in the recent and one Miocene species, these valves usually grow both upwards and downwards. Hence it would appear that there is some relation between the age of fossil Lepadidæ and the upward or downward direction of

[1] When I had the slices made, I did not know of H. von Meyer's paper on Aptychus, in the 'Acta Acad. Cæs. Leop. Car.,' vol. xv, Oct. 1829, tab. lviii and lix, fig. 13, in which perfectly accurate sections are given of the microscopical structure of *Aptychus lævis*.

the lines of growth in their valves. Aptychus, according to M. D'Orbigny, existed during the Carboniferous system, at a period vastly anterior to the oldest known Pollicipes, yet on the idea of its having been a Cirripede, the growth of its valves (Scuta) must have been upwards, as in the most recent forms; and it was allied to Lepas, that genus which, in the order of creation, and in the manner of growth, stands at the opposite end of the series from Pollicipes. From the several reasons now given, it does not appear to me that Aptychus, until weightier evidence is adduced, can be safely admitted as a Cirripede.

Geological History.—No true Sessile Cirripede[1] has hitherto been found in any Secondary formation; considering that at the present time many species are attached to oceanic floating objects, that many others live in deep water in congregated masses, that their shells are not subject to decay, and that they are not likely to be overlooked when fossilized, this seems one of the cases in which negative evidence is of considerable value. Mr. Samuel Stutchbury, moreover, (to whom I am deeply indebted for much information, and the loan of his beautiful collection of recent species,) has assured me that vast numbers of fossil secondary corals have passed through his hands, and that he has carefully looked without success for those genera which commonly inhabit living corals. Sessile Cirripedes are first found in Eocene deposits, and subsequently, often in abundance, in the later Tertiary Formations. These Cirripedes now abound so under every zone, all over the world, that the present period will hereafter apparently have as good a claim to be called the age of Cirripedes, as the Palæozoic period has to be called the age of Trilobites. There is one *apparent* exception to the rule that Sessile Cirripedes are not found in Secondary formations, for I am enabled to announce that Mr. J. de C. Sowerby has in his collection a Verruca (= Clisia, Clytia, Creusia, Ochthosia) from our English chalk: but this genus, though hitherto included amongst the Sessile Cirripedes, must, when its whole organisation is taken into consideration, be ranked in a distinct family of equal value with the Balanidæ and Lepadidæ, but perhaps more nearly related to the latter than to the Sessile Cirripedes. Hence the presence of Verruca in the Chalk is no real exception to the rule that Sessile Cirripedes do not occur in Secondary formations; on the contrary, it harmonises with the law, that there is some relation between serial affinities of animals, and their first appearance on this earth.

The oldest known pedunculated Cirripede is a Pollicipes, discovered by Professor Buckman in the Stonesfield Slate in the Lower Oolite: two species of the same genus have been described by Mr. Morris from the Oxford Clay, in the middle Oolite. I have

[1] Dr. Petzholdt has described and figured (Jahrbuch, 1842, p. 403, tab. x), a *Balanus carbonaria* from the carboniferous system; but as neither the operculum, the structure of the shell, the number of the valves, nor their manner of growth, can be made out or are described, the evidence appears quite insufficient to admit the existence of this genus at so immensely a remote epoch. Bronn, in the 'Index Palæontologicus,' gives, under Tubicinella, a cretaceous species; I have unfortunately not been able to consult the original work cited.

not heard of any Cirripede having been as yet discovered in the Upper Oolite, or in the Wealden formation. During the deposition of the great Cretaceous System, the Lepadidæ arrived at their culminant point; there were then three genera, and at least thirty-two species, some occurring in every stage of this system. Besides the thirty-two certainly known cretaceous forms, and several other doubtful ones, I believe that very many more will yet be discovered; I infer this from the fact, that in almost every collection lent to me for examination, although very small, I have found some new species. I have three or four species from the Gault; from five to eight in the Lower Chalk, and from nine to twelve species in the Upper Chalk (not including the Faxoe, Scanian, and Maëstricht stage); and of these nine to twelve species, five have been found by one collector, Mr. Fitch, in one locality, namely near Norwich. In Scania M. Angelin has found no less than nine or ten species, all belonging to the upper or Maëstricht stage of the Chalk. These fossils, judging from the habits of recent species of the same genera, were probably attached to fixed, or nearly fixed, objects at the bottom of the sea. Now at the present day, of attached Pedunculata (reckoning even Crustacea and Echinidæ as fixed objects), the whole Mediterranean and New Zealand can boast each only of three species, in both cases including Alepas, which is destitute of calcified valves and therefore not likely to be fossilized; Australia has three species; Madeira has four species, including one with very small and imperfectly calcified valves; the great Phillipine Archipelago, however, has afforded, owing to the labours of Mr. Cuming, as many as five species, though including one with horny valves, and a Lithotrya which lives embedded on the beach. Therefore since we already have nine or ten fossil species from one locality, and from the same stage of the chalk, we may admit that the pedunculated Cirripedes arrived during the upper part of the Cretaceous system at their culminant point.

Although, for this family, the number of species were considerable during the Cretaceous period, the individuals were mostly rare. I infer this from the small number of specimens in all collections; for instance, Mr. Fitch, who has assiduously collected for twenty years in the chalk near Norwich, possesses in his entire collection only nine keel-valves of *Scalpellum maximum*, and six of *S. fossula;* he has two Scuta (and with regard to these valves, it must be remembered, that each individual had two) of *Pollicipes striatus*, two of *P. fallax*, and four of *P. Angelini*. This occasional want of a relation, within the same region, between the number of the species in any given genus, and of the individuals appertaining to such species, is a singular fact, and has been strongly insisted on by Dr. Hooker, in regard to the Coniferous trees of the southern hemisphere: one would naturally have expected, that where circumstances favoured the existence of numerous species of a genus, they would likewise have favoured the multiplication of the individuals in all or most of such species; but this, as we here see, has not always been the case.

In the Eocene, Miocene, and Pliocene Tertiary deposits, I know only of two species of Scalpellum, and two of Pollicipes, with indications of two or three other species, all distinct from recent forms. It is a rather singular fact, considering the present wide distribution of the genus Lepas or Anatifa, and the frequency of the individuals, that not a single valve known certainly[1] to belong to this genus, or to any of the closely-allied genera, has hitherto been found fossil.

The oldest known cirripede is, as we have seen, a Pollicipes from the Lower Oolite, and it does not differ conspicuously from some of the recent species of the same genus; so, again, the cretaceous *Scalpellum fossula,* and the eocene *S. quadratum* are certainly very nearly related to the recent *S. rutilum* (nov. spec.). *Loricula* alone is a genus perfectly distinct from all living Cirripedia; and I may here add that of the Tertiary Sessile Cirripedes, I have hitherto not seen a single new generic form. This persistence of the same genera is somewhat remarkable, considering that amongst ordinary Crustacea nearly all the Secondary species belong to extinct genera;[2] it should, however, be borne in mind that Limulus has survived from the Palæozoic period to the present day. The Oolitic, Cretaceous, Tertiary, and recent species of Lepadidæ are all different from each other. By looking at the annexed Table, and putting out of question the species of which the age is uncertain, we have five common to two stages of the chalk; that is assuming for the present that the classification of the stages of the chalk commonly used and here followed, is correct. *Pollicipes glaber* is common to three, and, I believe, to four stages. *Scalpellum arcuatum* occurs in the Chalk-marl, and upper Greensand, and therefore this species also extends through three stages; but there is a slight difference between the specimens from the upper and lower stages, which some authors might perhaps consider specific. If fossil cirripedia had, like most recent species, very wide horizontal or geographical ranges, then, in accordance with a law now generally admitted, a considerable vertical range in some of the species is not improbable.

I may here observe that I am assured by Professors Forchammer and Steenstrup, that the formations of Scania and Westphalia are equivalent to that of Faxoe; and hence to that of Maëstricht. I have called these formations the "*Maestricht formation,*" to distinguish them from the common upper or white Chalk.

[1] In a mere catalogue, published without descriptions, in the 'Jahrbuch' for 1831, p. 155, by Hoenninghaus, *Anatifa cancellata* is given as a tertiary species: Mr. G. B. Sowerby has stated, in his 'Genera of Shells,' that he has seen a Tertiary specimen of this genus, but he cannot remember which valve it was.

[2] Pictet, Traité Élémentaire de Paléontologie, tom. iv, p. 4.

TABLE OF THE DISTRIBUTION OF THE SPECIES.

	Tertiary.	Faxoe, Scania, Maëstricht.	Upper Chalk.	Lower Chalk.	Chalk Marl.	Upper Greensand.	Gault.	Lower Greensand.	Middle Oolite.	Lower Oolite.
Scalpellum magnum . . .	*									
— quadratum . .	*									
— fossula	—	—	*							
— maximum . . .	—	*	*							
— lineatum . . .	—	—	—	*						
— hastatum . . .	—	—	—	—	*					
— angustum . . .	—	—	*?	*?	*?					
— quadricarinatum .	—	—	—	—	*					
— trilineatum . .	—	—	—	—	*					
— simplex . . .	—	—	—	—	—	—	—	*		
— arcuatum . . .	—	—	—	—	*	—	*			
— tuberculatum . .	—	—	*?	*?	*?					
— solidulum . . .	—	*								
— semiporcatum .	—	*								
— (?) cretæ	—	—	*							
Pollicipes concinnus . . .	—	—	—	—	—	—	—	—	*	
— ooliticus . . .	—	—	—	—	—	—	—	—	—	*
— Nilssonii . . .	—	*								
— Hausmanni . .	—	—	—	—	—	—	—	*		
— politus	—	—	—	—	—	—	*?			
— elongatus . . .	—	—	*							
— acuminatus . .	—	—	—	*						
— Angelini . . .	—	*	*							
— reflexus . . .	*									
— carinatus . . .	*									
— glaber	—	*?	*	*	*					
— unguis	—	—	—	—	—	—	*	*		
— validus	—	*								
— gracilis	—	—	*	*						
— dorsatus . . .	—	*								
— striatus	—	—	*							
— semilatus . . .	—	—	*?	*?	*?					
— rigidus	—	—	—	—	—	—	*			
— fallax	—	*	*							
— elegans	—	*								
— Bronnii	—	—	—	—	—	*				
— planulatus . . .	—	—	—	—	—	—	—	—	*	
Loricula pulchella	—	—	—	*						
Total 38	4	9–10	9–12	5–8	5–8	1	3–4	3	2	1

NOMENCLATURE OF THE VALVES.

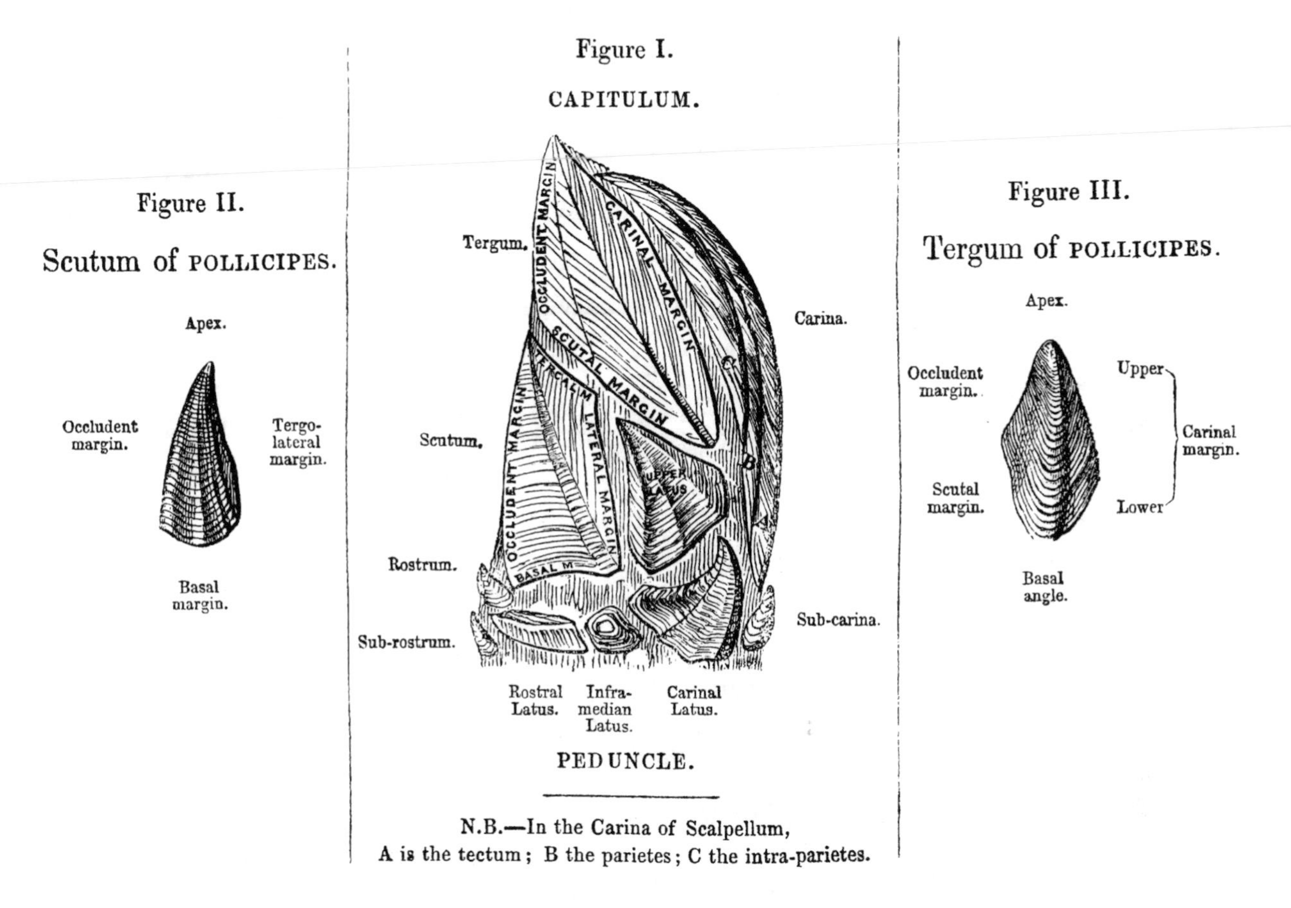

Figure I. CAPITULUM. PEDUNCLE.

N.B.—In the Carina of Scalpellum, A is the tectum; B the parietes; C the intra-parietes.

Figure II. Scutum of POLLICIPES.

Figure III. Tergum of POLLICIPES.

Whoever will refer to the published descriptions of recent and fossil Cirripedia, will find the utmost confusion in the names given to the several valves; thus, the valve named in the above woodcut, the Scutum, has been designated by various well-known naturalists as the "ventral," the "anterior," the "inferior," the "ante-lateral," and the "latero-inferior" valve; the first two of these titles have, moreover, been applied to the rostrum or rostral valve of Sessile Cirripedes. The Tergum has been called the "dorsal," the "posterior," the "superior," the "central," the "terminal," the "postero-lateral," and the "latero-superior" valve. The Carina has received the first two of these identical epithets, viz. the "dorsal" and the "posterior;" and likewise has been called the "keel-valve." The confusion, however, becomes far worse, when any individual valve is described, for the very same margin which is anterior or inferior in the eyes of one author, is the posterior or superior in those of another; it has often happened to me that I have been quite unable even to conjecture to which margin or part of a valve an author was referring. Moreover, the length of these double titles is inconvenient.

Hence, as I intend to describe all the recent and fossil species, I have thought myself justified in giving short names to each of the more important valves, these being common to the Pedunculated and Sessile Cirripedes.

The title of peduncle, which is either naked or squamiferous, requires no explanation; the scales and lower valves are arranged in whorls, which I have called by the botanical term of Verticillus. The part supported by the peduncle, and which is generally, though not always, in recent species protected by valves, I have designated the Capitulum.

I have applied the term *Scutum* to the most important and persistent of the valves, and which can almost always be recognised by the hollow giving attachment to the adductor scutorum muscle, from the resemblance which the two valves taken together bear to a shield, and from their office of protecting the front side of the body. From the protection afforded by the two *Terga* to the dorso-lateral surface of the animal, these valves have been thus called. The term *Carina* is a mere translation of the name already used by some authors, of Keel-Valve: in the genus Scalpellum, in which this valve is taken as typical, I have found it quite necessary, with fossil specimens, to distinguish the roof (see Woodcut, I,) or exterior surface, as the tectum (A); the inflected sides, as the parietes (B); and in several species in the upper half of the valve, the intra-parietes (C): the expressions of apex, basal margin, and inner margin, as applied to the Carina, require no explanation. The rostrum has been so called from its relative position to the Carina or keel. There is often a *sub-carina* and a *sub-rostrum*.

The remaining valves have been called *Latera;* there is always one large upper one inserted between the lower halves of the Scuta and Terga, and this I have named the Upper Latus or Latera; the other Latera in Pollicipes are numerous, and require no special names; in Scalpellum, where there are at most only three pair beneath the Upper Latera, it is convenient to speak of them (*vide* Woodcut, I,) as the *Carinal, Infra-median,* and *Rostral* Latera.

As each valve, especially amongst the fossil species, requires a distinct description, I have found it indispensable to give names to each margin. These have mostly been taken from the name of the adjoining valve, (see Woodcut, I.) In Pollicipes the margin of the Scutum adjoining the Tergum and Upper Latus, is not divided (Woodcut, II,) into two distinct lines, as in Scalpellum, and is therefore called the tergo-lateral margin; a narrow portion or slip along this side of the valve may be seen (Woodcut, II,) to be formed of upturned lines of growth; this is often of service in classification, and I have called it the tergo-lateral slip or segmentum tergo-laterale. In Scalpellum (Woodcut, I,) these two margins are separately named Tergal and Lateral. The angle formed by the meeting of the basal and lateral or tergo-lateral margins, I call the baso-lateral angle; that formed by the basal and occludent margins, I call, from its closeness to the Rostrum, the rostral angle. In Pollicipes the Carinal margin of the Tergum (Woodcut, III,) can be divided into an upper and lower Carinal margin.

That margin in the Scuta and Terga which opens and *shuts* for the exsertion and retraction of the cirri, I have called the Occludent margin.

During the periodical growth of the valves, especially when they are thick and massive, it happens in several species that the underlying corium deserts their upper ends or umbones, which consequently become marked by lines or ridges of growth, as I have called them, though perhaps lines of recession would have been more strictly correct. Such valves, consequently, have their upper ends projecting from and beyond the capitulum, and are said to project freely or *liberè;* this is often more especially the case with the Carina in Pollicipes, and in a lesser degree with the Terga.

From the peculiar curved position which the animal's body occupies within the capitulum, I have found it far more convenient (not to mention the confusion of nomenclature already existing) to apply the term Rostral instead of ventral, and Carinal instead of dorsal, to almost all the external and internal parts of the animal. Cirripedes have generally been figured with their surfaces of attachment downwards, hence I have termed the lower margins and angles the Basal, and those pointing in an opposite direction the Upper; strictly speaking, the exact centre of the usually broad and flat surface of attachment is the anterior end of the animal, and the upper tips of the Terga, the posterior end of that part of the animal which is externally visible; but in some cases, for instance in Coronula, where the base is *deeply concave,* and where the width of the shell far exceeds the depth, it seemed almost ridiculous to call this, the anterior extremity; as likewise does it in Balanus to call the united tips of the Terga, lying deeply within the shell, the most posterior point of the animal as seen externally.

CLASS—CRUSTACEA. SUB-CLASS—CIRRIPEDIA.

Family—LEPADIDÆ.

Cirripedia pedunculo flexili, musculis instructo: Scutis[1] *musculo adductore solummodŏ instructis: valvis cæteris, siquæ adsunt, in annulum immobilem haud conjunctis.*

Cirripedia having a peduncle, flexible, and provided with muscles. Scuta[1] furnished only with an adductor muscle: other valves, when present, not united into an immovable ring.

Besides the brief characters here given others might have been added, drawn from the softer parts of the animals, but as this Volume treats only of Fossil species, they would have been in this place superfluous. Nor have I thought it advisable to give here any definition of the Sub-class Cirripedia, or of the Order which contains both the Lepadidæ and Balanidæ, that is the Pedunculated and Sessile Cirripedes; for the characters would likewise have had to be derived almost entirely from the softer parts of the animal. It may, however, be worth stating, that by following the metamorphoses of the Cirripedia, it can be clearly shown that the capitulum together with the peduncle, in the Pedunculated Cirripedes, and that the shell together with the operculum in the Sessile Cirripedes, that is the whole of what is externally visible, consists simply of the first three segments of the head. In many Crustacea the carapace, formed by the backward production of the three anterior rings of the head, covers the dorsal surface of the thorax, and in some it encloses the limbs and mouth. This is likewise the case with the Cirripedia, and it is only the wonderful elongation of the anterior part of the head, its fixed condition, and the absence of external eyes and antennæ, which gives to the Cirripedia their peculiar character, and has hitherto prevented the homologies of these parts from having been recognised.[2]

[1] The meaning of this and all other terms is given in the Introduction at page 9.

[2] Nevertheless, in some Stomapoda, more especially in Leucifer of Vaughan Thompson, the anterior part of the head is only a little less elongated, compared with the rest of the body, than in the Cirripedia. That accomplished naturalist, M. J. D. Dana (Silliman's 'American Journal,' March, 1846,) has stated that "the pedicel of Anatifa corresponds to a pair of antennæ in the young:" although the peduncle or pedicel is undoubtedly thus terminated, this view cannot, I think, be admitted. In the larva, the part anterior to the mouth is as large, in proportion to the rest of the body, as in some mature Cirripedia: this anterior part supports only the eyes, antennæ, and two small cavities furnished with large nerves, which I

I may further state, that in the several Orders of Cirripedia such important differences of structure are presented, that there is scarcely more than one great character by which all Cirripedia may be distinguished from other Crustacea: this character is, that they are attached to some foreign object by a tissue or secretion (for at present I hardly know which to call it), which debouches, in the first instance, through the prehensile antennæ of the larva, the antennæ being thus embedded and preserved in the centre of the basis. The cementing substance is brought to its point of debouchement by a duct, leading from a gland, which (and this is perhaps the most remarkable point in the natural history of the Class) is part of and continuous with the branching ovaria. When we look at a Cirripede, we, in fact, see only a Crustacean, with the first three segments of its head much developed and enclosing the rest of the body, and with the anterior end of this metamorphosed head fixed by a most peculiar substance, homologically connected with the generative system, to a rock or other surface of attachment.

Genus—SCALPELLUM.

SCALPELLUM. *Leach.* Journ. de Physique, t. lxxxv, July, 1817.
LEPAS. *Linn.* Systema Naturæ, 1767.
POLLICIPES. *Lamarck.* Animaux sans Vertebres.
POLYLEPAS. *De Blainville.* Dict. des Sc. Nat., 1824.
SMILIUM (pars generis). *Leach.* Zoolog. Journal, Vol. 2, July, 1825.
CALANTICA (pars generis). *J. E. Gray.* Annals of Philosophy, vol. x, (2d series,) Aug. 1825.
THALIELLA (pars generis). *J. E. Gray.* Proc. Zoolog. Soc., 1848.
ANATIFA. *Quoy* et *Gaimard,* Voyage de l'Astrolabe, 1826—34.
XIPHIDIUM (pars generis). *Dixon.* Geology of Suffolk, 1850.

Valvis 12 *ad* 15: *Lateribus verticelli inferioris quatuor val sex, lineis incrementi plerumque convergentibus; Subrostrum rarissime adest: Pedunculo squamifero, rarissime nudo.*

suspect to be auditory organs; this part, therefore, I think, must unquestionably consist of the first two or three segments of the head: within it, even before the larva moults, the incipient striæless muscles and ovaria of the peduncle can be distinctly traced: immediately after the moult, we see this anterior part converted into a perfect peduncle; and for some time afterwards certain coloured marks, indicating the former position of the (so called) olfactory cavities and of the cast-off compound eyes, are still preserved. The prehensile antennæ are not cast off, for they are fastened down by the cementing substance, and are thus preserved in a functionless condition, with their muscles absorbed; after a time even the corium is withdrawn from within them. From the above and other coloured marks, and from the antennæ being preserved, it is easy to point out, in the peduncle of a young though perfect Lepas, the exact point which each part occupied in the head of the natatory larva.

Since the above was written, I find that Lovén has taken the same view of the homologies of the external parts of the Cirripedia; in his description of his *Alepas squalicola,* (Ofversigt of Kongl. Vetens., &c., Stockholm, 1844, pp. 192—4,) he uses the following words: "Capitis reliquæ partes, ut in Lepadibus semper, in *pedunculum mutatæ et involucrum,*" &c.; his involucrum is the same as the Capitulum of this work.

CHARACTERES VALVARUM IN SPECIEBUS FOSSILIBUS.

Carina angusta, introrsùm arcuata, ab apice ad marginem basalem paululum dilatata; parietes valde inflexi, costis manifestis a tecto plerumque disjuncti; in multis speciebus intra-parietibus instructi: intra-parietes nonnunquam supernè producti ultra Umbonem, qui fit inde subcentralis: parietum lineæ incrementi perobliquæ. Scuta plerumque subconvexa et tenuia, trapezoidea; marginibus tergalibus lateralibusque angulo insigni disjunctis.

Sect. †. *Subcarina adest (solummodò species recentes).*

Sect. ††. *Subcarina deest.*

A. Valvæ quatuordecim: Carinæ umbone subcentrali.

B. Valvæ duodecim: Carinæ umbone ad apicem posito.

Valves 12 to 15 in number. Latera of the lower whorl, four or six, with their lines of growth generally directed towards each other. Sub-rostrum[1] very rarely present. Peduncle squamiferous, most rarely naked.

CHARACTERS OF THE VALVES IN FOSSIL SPECIES.

Carina narrow, bowed inwards, widening but little from the apex to the basal margin, having parietes much inflected, and generally separated by distinct ridges from the tectum, and having in many species intra-parietes, which are sometimes produced upwards beyond the umbo, so as to make it sub-central; lines of growth on the parietes very oblique. Scuta generally only slightly convex and thin, four-sided, the tergal and lateral margins distinctly separated by an angle.

Sect. †. Subcarina present. (This section includes only recent species.)

Sect. ††. Subcarina absent.

A. Valves fourteen in number; Carina with the umbo subcentral.

B. Valves twelve; Carina with the umbo at the apex.

The first of the above two paragraphs contains the true Generic description (here leaving out the softer parts), as applicable to recent and, as far as known, to fossil species: the second paragraph has been drawn up to aid any one in classifying the characteristic valves, when found separated, as is most frequently the case with all fossil Pedunculata. The first or proper Generic characters would have been more precise, had it not been for the existence of one recent species, the *S. villosum* (*Pollicipes villosus*, Leach, *Calentica Homii*, J. E. Gray,) which leads into the next genus Pollicipes. I mention this species in order to confess, that had the valves been found separately, and their number unknown, they would certainly have been included by me under Pollicipes, although, taking the whole organisation into consideration, I have determined to include this species under Scalpellum. I need not

[1] The meaning of this and all other special terms is given in the Introduction at p. 9.

here repeat the remarks made in the Introduction on the great difficulties in classifying the recent species, and still more the fossil species of Scalpellum. I may, however, here state that should the *S. vulgare* be hereafter kept distinct in a genus to itself, *S. magnum* would have to go with it. Should a recent species, which in a future work I shall describe under the name of *S. rutilum*, be generically separated, it will probably have to bear the name of Xiphidium, from its alliance to the Eocene *X. quadratum* of Sowerby, to which species the cretaceous *S. fossula* and several other forms are apparently closely allied. These latter species, however, are likewise closely allied to the *Scalpellum ornatum*, which Mr. Gray has already raised to the rank of a genus under the name of *Thaliella*. There are some fossil species, as *S. arcuatum*, and *simplex* and *solidulum*, which I cannot rank particularly near any recent forms. Mr. Sowerby founded the genus Xiphidium on the umbo in the Carina being situated at the apex, and on its growth being consequently exclusively downwards. This is likewise the case with the recent *S. rutilum* ; but I shall have occasion to show, under *S. magnum*, that the upward growth of the Carina in that and other species of the genus, depends merely on the intra-parietes, which are present in many species, meeting each other and being thus produced upwards. Moreover, in the recent *S. ornatum*, the position of the umbo is variable, according to the age of the specimen; in half-grown individuals being seated at the apex, and in large specimens being sub-central, as in *S. vulgare, magnum*, and other species. I should have been very glad to have retained the genus Xiphidium, but taking into consideration the whole organisation of the six recent species, I can only repeat that we must either make six genera of them, or leave them altogether, and this latter has appeared to me the most advisable course.

Sexual Peculiarities.—For reasons stated in the Introduction, I have kept the genera Scalpellum and Pollicipes distinct; but I may mention, in order to call attention to a point of structure which may hereafter be discovered in some fossil species, that I was much influenced in this decision by some truly extraordinary sexual peculiarities in all six recent species of Scalpellum. *Scalpellum ornatum* is bisexual; the individual forming the ordinary shell, is female; each female has two males (a case of *Diandria monogynia*), which are lodged in small transverse depressions, one on each side, hollowed out, on the inner sides of the Scuta, close above the slight depressions for the adductor scutorum muscle; in *S. rutilum* (nov. spec.) two males are lodged in the same place on each side, but rather in concavities in the valve, than in distinct depressions. As these are the two recent species most nearly related to several Cretaceous and Eocene forms, we might expect to find similar depressions in some fossil species; but as yet I have not succeeded in distinctly finding such. The male cirripedes are very singular bodies; they are minute, of the same size as the full-grown larva; they are sack-formed, with four bead-like rudimental valves at their upper ends; they have a conspicuous internal eye; they are absolutely destitute of a mouth, or stomach, or anus: the cirri are rudimental and furnished

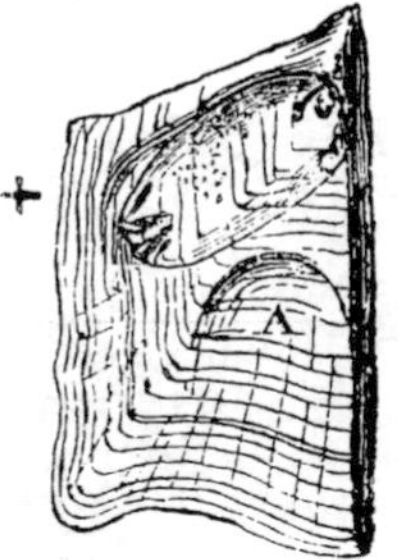

Inside view of the Scutum in *Scalpellum ornatum*. (A) is the depression for the adductor muscle.

with straight spines, serving, apparently, to protect the entrance of the sack: the whole animal is attached, like ordinary cirripedes, first by the prehensile antennæ, and afterwards by the cementing substance; the whole animal may be said to consist of one great sperm-receptacle, charged with spermatazoa; as soon as these are discharged, the animal dies.

A far more singular fact remains to be told: *Scalpellum vulgare* is like ordinary cirripedes, hermaphrodite, but the male organs are somewhat less developed than is usual; and, as if in compensation, several short-lived males are almost invariably attached on the occludent margin of both Scuta, at a spot marked by a fold (not thus caused), as may be seen on the inside view of this valve in the fossil *S. magnum*, which, in all probability, was furnished with them. I have called these beings complemental males, to signify that they are complemental to an hermaphrodite, and that they do not pair like ordinary males with simple females. In *Scalpellum vulgare*, the complemental male presents only slight specific differences from the male of *S. ornatum*. It would be foreign to the purpose of this volume here to enter on further details; nor should I have touched on the subject, had I not wished specially to call attention to the presence of cavities on the under sides of the Scuta above the pits for the adductor muscle. I will only add, that in the other species of Scalpellum, the complemental males are more highly organised, and are furnished with a mouth and prehensile cirri; the valves are more or less rudimental in the different species; these complemental males are not always present, and are never attached to young hermaphrodites; when present, they adhere in such a position, that they can discharge their spermatozoa into the sack of the hermaphrodite: their attachment does not affect the form of the valves.[1]

Description of Valves.—It will, I think, be most convenient to confine the following description to the fossil species of the genus. No one specimen has been found quite perfect; but, judging from analogy, the capitulum was probably formed of fourteen valves in *S. magnum*, and of twelve in the remaining species. These valves are commonly smooth,

[1] Exactly analogous facts are presented, though more conspicuously, by the two species of the genus *Ibla*. Before examining this genus, I had noticed the complemental males on *Scalpellum vulgare*, but had not imagined even that they were Cirripedia. *Ibla Cumingii* (as I propose to call a new species collected by Mr. Cuming, at the Philippines) is bisexual; one or two males being parasitic near the bottom of the sack of the female. These males are small, are supported on a long peduncle, but are not enclosed in a capitulum (such protection being here unnecessary), are furnished with a mouth, ordinary trophi, stomach, and anus; there are only two pair of cirri, and these are distorted, useless and rudimentary; the whole thorax is extremely small; there is no penis, but a mere orifice beneath the anus for the emission of semen: hence *Ibla Cumingii* is exactly analogous to *Scalpellum ornatum*. On the other hand, the closely allied Australian *Ibla Cuvierii*, like *Scalpellum vulgare*, is hermaphrodite, but has, in every specimen opened by me, a complemental male attached to near the bottom of the sack; this complemental male differs only about as much from the male of *Ibla Cumingii*, as the female *J. Cumingii* differs from the hermaphrodite form of *I. Cuvierii*. I intend hereafter to give detailed anatomical descriptions and drawings of the males and complemental males of Ibla and Scalpellum.

but in two or three species are marked with longitudinal ridges; they are generally rather thin; this, however, is a character which is variable even in the same species.

Carina narrow, widening but little from the apex downwards, slightly or considerably curved inwards, with the umbo seated at the uppermost point: *S. magnum*, however, must be excepted, for in it the umbo is sub-central, and the valve almost angularly bent, as will be described in detail under that species. The apex rarely projects freely; but this is a variable point in the same species; the basal margin is either pointed, rounded, or rarely truncated. The chief character by which this valve can be recognised, as belonging to the genus Scalpellum, is the distinct separation by an angle, (see woodcut, Fig. 1, in the Introduction,) often surmounted by a prominent ridge, of the tectum or roof, from the parietes, which are either steeply or rectangularly inflected; the lines of growth on these parietes are oblique. A still more conspicuous character is afforded by the part (when present), which I have called the intra-parietes; these give to the valve a pieced appearance, and seem let in, to fill up a vacuity between the *upper part* of the carina and the terga, and this is their real office; they are separated from the true parietes by a ridge, which evidently marks the normal outline of the valve. These intra-parietes are flat, and they have a striated appearance rather different from the rest of the valve; and the lines of growth on them are extremely oblique, almost parallel to the inner margins of the valve.

Scuta very slightly convex; four-sided; the tergal and lateral margins being divided by a slightly projecting point or angle; and this is the chief character by which the scuta of this genus can be distinguished from those of Pollicipes. The umbo is seated at the uppermost point, except in *S. magnum*, and in *S.* (?) *cretæ* (Tab. I, fig. 1 *c*, and fig. 11 *c*), in which species the lines of growth, instead of terminating at the angle separating the lateral and tergal margins, are produced upwards, so that the valve is added to above the original umbo. In *S. tuberculatum* (fig. 10 *d*), the scuta present an intermediate character between that in ordinary fossil species, for instance in *S. fossula* (fig. 4 *a*), and in *S. magnum* and *cretæ*. The occludent margin is nearly straight, or slightly curved; both it and the lateral margin form nearly rectangles with the basal margin, which is nearly straight. Internally the depression for the adductor scutorum is generally, but not always, very plain; sometimes the valve is filled up and rendered solid in the upper part above the adductor muscle. The apex sometimes projects freely, and is internally marked with lines of growth. The internal occludent margin, or edge, is also often marked by lines of growth, and the part thus marked, close above the adductor muscle, sometimes becomes suddenly wider; this is caused by some slight change in the position of the animal's body during growth.

Terga flat, either trigonal or rhomboidal, and, in the former case, sometimes so much elongated, with the carinal margin so much hollowed out, as to become almost crescent-shaped; a slight furrow often runs from the upper to the basal angle. Internally, in the upper part, there is in some species a little group of small longitudinal ridges, unlike anything I have seen in recent species, and serving, I apprehend, to give firmer attachment to the corium.

Rostrum unknown in any fossil species; but judging from recent species, it probably existed in all.

Upper latera known only in three species; in *S. magnum* it is irregularly oval, with the umbo central: in *S. quadratum* and *fossula*, five-sided, with the umbo at the upper angle: in the eocene *S. quadratum*, however, an inner ledge very slightly projects beyond the two upper sides, and first indicates a tendency to upward growth. *Rostral latera*, known only in *S. magnum* and *quadratum*, they are transversely elongated, narrow, and small. *Infra-median latera* unknown; they probably existed only in *S. magnum*. *Carinal latera*, known in *S. magnum*, *quadratum*, *fossula*, *solidulum*, and *maximum*; in the first species they are transversely elongated; in the three latter, of an irregular curved shape, and flat. In the fossil and recent species, the rostral and carinal latera grow chiefly in a direction towards each other; so that their umbones are close to, or even seated exteriorly to, the carinal and rostral ends of the capitulum. *Peduncle*, calcified scales are known only in one species, the *S. quadratum*; but they probably existed in all: the naked peduncle, however, of the recent *S. Peronii* must make us cautious on this head.

[*A*] *Valvæ quatuordecem: Carinæ umbone sub-centrali.*

1. Scalpellum magnum.[1] Tab. I, Fig. 1.

S. Laterum carinalium et rostralium umbonibus liberè (sicut cornua) prominentibus, dimidiam seu tertiam partem longitudinis valvarum æquantibus.

Carinal and rostral latera, with their umbones projecting freely like horns, and equalling one half or one third of the entire length of these valves.

Coralline Crag (lower part). Sutton, Gedgrave, Sudbourne. *Mus.* S. Wood and Lyell.

From the close affinity between this species and the recent *Scalpellum vulgare*, we may confidently infer that the capitulum consisted of fourteen valves, which are all preserved in Mr. Wood's collection, with the exception of the infra-median latera and of the rostrum. This latter valve would, no doubt, be rudimentary, and it has been overlooked by naturalists even in the recent species. The chief difference, excepting size, between these two species, is in the form of the rostral and carinal latera, but unfortunately these valves are extremely variable. It might even be maintained, with some degree of probability, that *S. magnum* was only a variety of *S. vulgare*. The valves of *S. magnum* are all thicker, stronger, more rugged, and considerably larger than in *S. vulgare*. Taking

[1] I have followed Mr. Morris in his Catalogue, in adopting this name from the MS. of Mr. Searles Wood, to whose kindness I am greatly indebted for having placed in my hands the whole of his large series of valves of this species.

the largest scutum, tergum, carina and upper latera in Mr. Wood's collection, they are very nearly double the size of the same valves in the largest specimen of *S. vulgare* seen by me, namely from near Naples, which had a capitulum eight tenths of an inch in length; and they are more than double the size of the same valves in any British specimen. *Scalpellum magnum* probably had a capitulum one inch and a half in length.

Carina (Tab. I, fig. 1 *b* and *f*) abruptly, almost rectangularly bent, with the umbo of growth seated just above the bend, at about one third or one fourth of the entire length of the valve from the upper point; form linear, with the lower part slightly wider than the upper. Exteriorly the surface is rounded with no central ridge, excepting near the umbo, where the narrowness of the whole valve gives it a carinated appearance; basal margin rounded. From the umbo two faint ridges run to each corner of the basal margin, separating the steeply-inclined parietes from the roof,—a character of some importance in the cretaceous species of this genus: outside of these two ridges there are other two ridges, not extending down to the basal margin, and separating the parietes from the intra-parietes, which latter being united at their upper ends, and produced upwards, form that part of the carina which is above the umbo. By comparing the lateral views of the carina of the cretaceous *S. fossula* (fig. 4 *c*), and of this species, it will be seen, that the apparently great difference of the umbo of growth being either at the apex, or, as in this species, sub-central, simply results from the lines of growth of the intra-parietes meeting each other, the valve being thus added to at its upper end. The carina of *S. magnum*, examined internally, is found often to be narrower under the umbo than either above or below it, a character I have not seen in the recent *S. vulgare*. The lateral width or depth of the valve (measured from the umbo to the inner edge) is also greater than in *S. vulgare*: this portion is internally filled up and solidified. No part of the apex of the valve projected freely. The longest perfect specimen which I have seen, is half an inch in length; but I have noticed fragments indicating even a greater size.

Scuta (fig. 1 *c*) much elongated, trapezoidal, slightly convex; umbo placed on the occludent margin at about one fourth of the entire length of the valve from the apex, so that the valve grows upwards and downwards. Occludent margin straight, slightly hollowed out above the umbo, forming rather less than a right angle with the basal margin, which latter is at right angles to the lateral margin. The tergal margin is separated from the lateral by a slight projection (beneath which the margin is a little hollowed out), and from this projection there runs a ridge, often very conspicuous, to the umbo. The part above the ridge, stands at rather a lower level than that below it, and the lines of growth on it are generally less distinct. This is connected with the fact, as ascertained in *S. vulgare*, that the valve, during its earliest stage, grows only downwards, the ridge thus indicating the original form of the valve and tendency of the lines of growth. On comparing that part of the scuta beneath the umbo and ridge, in the present species (Tab. I, fig. 1 *c*), with the whole valve in some other species, for instance in *S. fossula* (fig. 4 *a*), in which the umbo is seated at the apex, as it was in the first commencement of growth in *S. vulgare* and *magnum*, it will be seen how closely the two valves resemble each other. The scutum of *S. tuberculatum* (fig. 10 *d*) is intermediate in its manner of growth between those of *S. magnum* and *fossula.* Internally, the impression for the adductor muscle is deep: on the occludent margin, close to the umbo, there is a deep fold, which is connected with the growth of the upper part of the valve being subsequent to that of the lower part. There is very little difference between this valve and that of *S. vulgare;* the upper part, however, appears to be always thicker. Length of largest specimen one eighth of an inch.

Terga (fig. 1 *d*) triangular, sometimes approaching to crescent-shaped; flat and thin, though the thickness of the valve varies. Carinal margin straight, or very slightly hollowed out; in its upper part there is a barely perceptible prominence marking the limit of the upward extension of the carina. Basal angle blunt, rounded; from it a line, formed by the convergence of the zones of growth, runs near and parallel to the carinal margin, up to the apex. Occludent margin about equal in length to the scutal; parallel to the former, a slip of the valve is rounded and slightly protuberant, and this portion projects a little on the scutal margin. A very small portion, or none, of the apex of the valve projected freely. This valve is somewhat narrower, and the scutal margin straighter than in *S. vulgare.*

Rostrum unknown, no doubt rudimentary, probably quadrangular.

Upper latera (fig. 1 *e*) flat, oval, with the upper half a little pointed; the lower margin shows traces in a varying degree consisting of three sides. The surface, but chiefly of the lower half, is faintly marked with striæ radiating from the centre. The umbo lies in the middle, and from it two slight ridges, first bending down, diverge on each side. In *Scalpellum vulgare* this valve (which is very similar in shape to that of *S. magnum*) at the first commencement of its growth, as with the scuta, is added to only downwards; and thus the two diverging ridges mark the form which the valve originally tended to assume: bearing in mind that the basal margin tends to be three sided, if we remove that part of the valve above the ridges which have been superadded to the original form, we shall have a five-sided valve, essentially like that in the *S. quadratum* and *S. fossula* (fig. 3 *e*, and fig. 4 *d*).

Rostral latera (fig. 1, *g* to *k*) elongated, widening gradually from the umbo to the opposite end, which is equably rounded: umbonal half free, curling outwards; the internal surface of the other half (*h*) is nearly flat and regularly oval, with its end towards the umbo pointed; the freely projecting portion varies from nearly one half to one third of the entire length of the valve; but in one *distorted* specimen it was only one sixth of this length. The width, also, of the valve varies (*g* and *h*), compared to its length. This valve, compared with its homologue in *S. vulgare,* differs more than any of the preceding valves; it is proportionally larger, and the internal or growing surface is oval, instead of being oblong and almost quadrangular; and the umbonal or freely projecting portion in *S. vulgare* is only one sixth or one seventh of the entire length of the valve.

Infra-median latera unknown.

Carinal latera (fig. 1, *l* to *n*) narrow, thick, much elongated, widening gradually from the umbo to the opposite end, which is rounded and obliquely truncated. Surface, exteriorly flat; internally convex. The umbonal, freely projecting portion is sometimes more than half, sometimes only about one third, of the entire length of the valve. This portion curls outwards and likewise upwards. The degree of curvature and the width (*m* and *n*), in proportion to the length, varies. The upper and lower margins are approximately parallel to each other; the umbonal end of the growing surface is bluntly pointed. This valve differs from its homologue in *S. vulgare*, in being larger, much narrower in proportion to its length, more massive, and with a far larger portion of the umbonal end freely projecting; also in the approximate parallelism of the upper and lower margins, and in the umbonal end of the growing surface being pointed instead of square. In *S. vulgare* the upper margin is much more curled upwards than the lower, and the freely projecting portion is only one fifth of the entire length of the valve.

Taking the largest specimens in Mr. Wood's collection, the freely projecting portions of the carinal latera must have stuck out like horns, curling from each other and a little upwards, for a length of a quarter of an inch. So again, the much flattened horns of the rostral latera, curving from each other, but not upwards, must have projected half an inch beyond the probably rudimentary rostrum. The capitulum must have presented a singular appearance, represented in the imaginary restored figure (fig. 1 *a*), with its pair of projecting horns at both ends.

Peduncle; calcareous scales unknown, but undoubtedly they existed.

Varieties: the variation in the rostral and carinal latera has already been pointed out. In Mr. Wood's collection there are numerous scuta, terga, carinæ, and carinal latera, from Sutton; and these are all smaller than those above described, which come from Sudbourne, and than some others in Sir C. Lyell's collection from Gedgrave. All these places, however, belong (as I am informed by Mr. Wood) to the same stage of the Coralline Crag. In the Sutton specimens the carinal latera show the same character as in those from Sudbourne, but the carina apparently is not internally so much narrowed in under the umbo; this, however, is a character which is conspicuous only in the larger Sudbourne specimens, and anyhow cannot be considered as sufficient to be specific.

I may take this opportunity of stating, that in Mr. Harris's collection of organic remains from the chalk detritus, at Charing, in Kent, I have found the upper part of a carina of a very young and minute Scalpellum, which cannot be distinguished from this species; but considering the state of the specimen, it would be extremely rash to believe in their identity. All the known cretaceous species have the umbo at the apex, so that the Charing specimen differs remarkably from its cretaceous congeners.

[*B*] *Valvæ duodecem: Carinæ umbone ad apicem posito.*

2. SCALPELLUM QUADRATUM. Tab. I, fig. 3.

XIPHIDIUM QUADRATUM. *Dixon*, in Sowerby's Mineral. Conch., Tab. 648; Geology of Suffolk, Tab. xiv, figs. 3 and 4.
POLLICIPES — ? *J. Sowerby*. Geolog. Trans., 2d series, vol. v, pl. 8, fig. 5.

S. tecto parietibusque carinæ planis, lævibus, simplicibus, margine basali ferè rotundato; Lateribus superioribus quinque-làteralibus, lævibus.

Carina, with its tectum and parietes flat, smooth, and simple; basal margin almost rounded. Upper latera five-sided, smooth.

Eocene Tertiary. Bognor; Hampstead. *Mus.* S. Wood, F. Edwards, N. Wetherell.

My materials consist of a slab of rock, belonging to Mr. S. Wood, almost made up of the valves of this species, of two beautiful specimens in Mr. F. Edwards's collection, and of some excellent drawings from Mr. Dixon's specimens by Mr. James de C. Sowerby, in the Mineral Conchology.[1] The valves in several of these specimens are nearly in their proper positions, though there is not one in which they have not slipped a little. Their relative positions are given, I believe nearly correctly, in Pl. I, fig. 3 *a*. Their number I have little doubt was twelve. This, however, includes a rostrum, probably almost rudimentary, the existence of which I infer only from the analogy of all recent species. Mr. J. Sowerby supposed that there were, as in *S. vulgare*, four pair of latera (and therefore fourteen valves in all), but I conclude, without hesitation, that there were only three pair, as in the recent *S. rutilum* (*nov. spec.*), to which the *S. quadratum* is much more nearly allied than to *S. vulgare*.

Capitulum: elongated, probably composed of twelve valves. *Carina* (fig. 3, *d, i, k*), rather narrow, slightly and regularly bowed and widening from the apex to the basal margin, which latter is bluntly pointed, or almost rounded; internally deeply concave; externally with the tectum and parietes flat, and at right angles to each other;—hence the carina is square-edged, and its specific name has been given to it. *Scuta* (fig. 3, *b, h*) oblong, occludent margin slightly arched, forming with the basal rather less than a right angle; tergal margin separated by a just perceptibly projecting point from the lateral margin, which latter is very slightly hollowed out; whole valve slightly convex, with a trace of a ridge running from the apex to the baso-lateral angle. Internally (*h*), there is a large pit for the adductor scutorum, above which there is a slight depression or fold marked with curved lines of growth, and in this depression on each side complemental males

[1] Some small fragments were found by Mr. Wetherell, and are noticed in his Paper in the fifth volume of the 'Geolog. Transactions,' entitled "Observations on a Well dug on the south side of Hampstead Heath."

were probably attached. *Terga* (fig. 3 *c*) triangular, large, flat, basal angle bluntly pointed; apex slightly projecting, as a solid horn; occludent margin very slightly arched. *Rostrum* unknown; judging from the narrowness of the umbones of the rostral latera, it was probably minute or rudimentary. *Upper latera* (fig. 3 *e*) large compared with the lower valves, flat, five-sided, with the two upper sides the longest; of the three lower sides, that corresponding with the end of the rostral latera is generally (especially in young specimens) the shortest. Umbo seated at the uppermost angle; but in full-sized specimens, a narrow ledge has been added, during the thickening and growth of the valve, along the two upper margins, and consequently round the apex. *Rostral latera* (fig. 3 *f*) extremely narrow, three or four times as long as wide; considerably arched, extending parallel to the basal margin of the scuta; widening gradually from the umbo to the opposite end, which is obliquely truncated in a line (as I believe) corresponding with the shortest side of the upper latera; inner surface smoothly arched; during growth, the narrow rostral half of the valve becomes much thickened, and at the same time added to along its upper margin, thus producing a solid, sloping, projecting edge; umbo slightly projecting. *Carinal latera* (fig. 3 *g*) almost flat, not elongated, of a shape difficult to be described; approaching to a triangle, with curved sides, and one angle protuberant.

Peduncle. The calified scales are apparently large in proportion to the valves of the capitulum; transversely elongated, pointed at both ends, and more or less crescent shaped.

Affinities. This species was generically separated from Scalpellum by Mr. Dixon, as I am informed by Mr. James Sowerby, solely owing to the umbo of growth in the carina being at the apex, instead of being sub-central, as in *S. vulgare;* but I need not here repeat the reasons already assigned for at present keeping all the recent and fossil species under the same genus. In the umbo of growth, in the carina and scuta being seated at their upper ends, in the square form of the carina, in there being only three pair of latera, and in the large size of the upper latera, this eocene species is much more closely allied to *S. rutilum* (*nov. spec.*, of which the habitat is unfortunately not known,) than to any other recent species. In some respects, however, I may remark, *S. rutilum* is even more closely related to certain cretaceous forms. To *S. ornatum*, the nearest recent congener of *S. rutilum*, the present species is allied by the narrowness of the rostral latera, and by the large size and peculiar shape of the scales on the peduncle: the carinal latera perhaps rather more resemble those of *S. vulgare* than of any other *recent* species. Certainly, all the affinities in *S. quadratum* point to *S. rutilum*, *ornatum*, and *vulgare*, and these three recent species are characterised by having males or complemental males attached to the sides of the orifice of the sack, whereas, in the other species, they are elsewhere attached; hence it is that I believe that males were probably lodged in the slight depressions described on the inner sides of the scuta; but the depression is not here nearly so distinctly developed as it is in the recent *S. ornatum*, and more resembles the fold on the occludent edge of the valve in *S. vulgare:* I must add that folds of this nature do not necessarily imply the presence of males.

3. SCALPELLUM FOSSULA. Tab. I, fig. 4.

POLLICIPES MAXIMUS. *J. Sowerby.* Min. Conch. Tab. 606 (*a tergum*), fig. 3.

S. carinâ intra-parietibus instructâ; tecto utrinque costis magnis, tumidis, superne planatis, marginato; margine basali obtusè acuminato. Lateribus superioribus quinque-lateralibus; costis duabus modicis ab apice ad marginem basalem continuatis.

Carina, having intra-parietes, with the tectum bordered on each side by large, protuberant, flat-topped ridges; basal margin bluntly pointed; upper latera five sided, with two slight ridges extending from the apex to the basal margin.

Upper Chalk. Norwich; Northfleet, Kent. *Mus.* Fitch, J. de C. Sowerby, Wetherell.

General Remarks. My materials consist of two specimens, belonging to Mr. Fitch, most kindly lent me for examination; in which, taken together, the scuta, terga, carina, upper and carinal latera, are seen almost in their proper places. In Mr. J. Sowerby's collection there is a single scutum, also, from Norwich. From analogy with the eocene *S. quadratum* and the recent *S. rutilum,* I have little doubt that there were only three pair of latera; and that, probably, there was a rostrum. With respect to the *exact* position of the carinal latera, there is, as also in the case of the *S. quadratum,* some little doubt.

Capitulum narrow, elongated, probably composed of 12 valves, which are moderately strong, and apparently closely locked together. The length of the capitulum in the largest specimen was 1·1 of an inch.

Carina (fig. 4, *c, g, h*) strong, moderately bowed, extending far up between the terga, almost to their upper ends; rather narrow throughout, gradually widening from the apex to the base; lines of growth plain; no portion projects freely. The tectum or central portion is slightly arched, subcarinated, and bounded on each side by flat-topped, protuberant ridges: the tectum terminates downwards in a blunt point (the two margins forming an angle of rather above 90°), which projects beyond the bounding ridges; the tectum and the two bounding ridges all widen gradually from the apex towards the base. The parietes are channelled or concave; they do not extend so far down as the ridges bounding the tectum. In the upper half of the carina, we here first see the additional parietes, or intra-parietes, which appear as if formed subsequently to the other parts, and let in between the ordinary parietes of the carina, and the terga. It has been already shown, under *S. magnum,* that it is the intra-parietes produced upwards, which causes in that and some other species the umbo of the valve to be subcentral.

Scuta (fig. 4, *a,f*) oblong, the basal margin only slightly exceeding half the entire length of the valve; valve strong, rather plainly marked with lines of growth; basal margin at nearly right angles to the occludent margin; tergal margin separated by a slightly-projecting point from the lateral margin, which in the lower half is slightly protuberant; tergal margin straight, with the edge thickened and slightly reflexed. A distinct, square-edged ridge (therefore formed by two angles) runs from the umbo to the baso-lateral angle, which is itself obliquely truncated. Internally (*f*), there is a large and deep pit for the adductor scutorum. *Terga* (fig. 4 *b*) triangular, flat, large, fully one third longer than the scuta; basal half much produced; basal angle pointed; from it to the apex or umbo there runs a narrow, almost straight furrow, at which the lines of growth converge—it runs at about one third of the entire width of the tergum (in its broadest part) from the carinal margin. Parallel to the occludent margin, and at a little distance from it, there runs a wide, very shallow depression up to the apex. The scutal margin is not quite straight, about a third part, above a slight bend corresponding with the apex of the upper latera, being slightly hollowed: from the above bend a very faint ridge runs to the apex of the valve. *Upper latera* (fig. 4 *d*) large, flat, with five sides, of which the two upper are much the longest; the basal side is next in length, and the scutal side much the shortest. As far as I can judge of the positions of the lower valves, with respect to the upper latus, I believe, that the rostral latera, probably, abutted against the shortest of the three lower sides; that the carina ran along the one next in length, and the carinal latera along the middle basal side, which I suppose extended in an oblique line, and not parallel to the base of the capitulum: the two upper long sides no doubt touched the scuta and terga. The umbo of growth is at the apex; there is, however, a trace of a projecting ledge added round the upper margins during the thickening of this upper part of the valve. Two slight ridges run from the apex to the two corners of the middle of the three lower sides. *Carinal latera* (fig. 4 *e*): these are not quite perfectly seen: the umbo forms a sharp point, whence the valve rapidly expands and curves apparently downwards and towards the upper latera. Near one margin there is a very narrow furrow, and on the other a wide depression, both running and widening from the umbo to the opposite end, which is slightly sinuous. I imagine these carinal latera occupied a nearly triangular space between the middle of the three lower sides of the upper latera and the basal portion of the carina. *Rostral latera, rostrum* and *peduncle* unknown; the *rostral latera* must have been very narrow.

Affinities.—In the shape and manner of growth of the scuta, and more especially of the upper latera, this species is certainly more closely allied to the eocene *S. quadratum*, than to any other species; but in the peculiar characters of the carina, it is nearer to the recent *S. rutilum*; we have previously seen that the nearest congener to *S. quadratum* is this same *S. rutilum*. The most conspicuous diagnostic character of this species is derived from the peculiar form of the carina,—its tectum being bounded by a rounded ridge on each side. The square-edged ridge running from the apex to the baso-lateral angle of the scuta is a trifling, but I believe, a diagnostic character. If I am right in placing *S. rutilum* in the genus Scalpellum, and I think there can be no doubt of this, considering the characters of its complemental male, then there can be no question that the present species belongs to the same genus.

4. Scalpellum maximum, Tab. II. fig. 1—10.

Pollicipes maximus. *J. Sowerby.* Min. Conch., tab. 606, solummodo, fig. 4 et fig. 6. N. B.—Fig. 3 est Tergum *S. fossulæ,* et fig. 5 alia species ignota.
— maximus. *Steenstrup.* Kroyer Tidsskrift, b. ii, pl. v, figs. 17, 18.
— medius. *Steenstrup.* Kroyer Tidsskrift, b. ii, pl. v, figs. 13, 13*, 33.
— sulcatus. *J. Sowerby.* Min. Conch., pl. 606, fig. 2, sed non fig. 1 et 7.

S. carinâ intra-parietibus instructâ; tecto subangulari vel subcarinato; margine basali rectangulariter acuto; totâ valvâ plus minusve introrsùm arcuatâ, sed margine interno ferè recto; tecto[1] *transversè plus minusve convexo; superficie pænè lævi, striis paucis obsoletis longitudinalibus elevatis; tectum, parietes, et intra-parietes inter se separantur costis plus minusve prominentibus.*

Carina having intra-parietes, with the tectum slightly angular or subcarinated, basal margin rectangularly pointed: whole valve more or less bowed inwards, but with the inner margin nearly straight; tectum, in a transverse line, more or less convex; surface nearly smooth, with a few faint longitudinal raised striæ; more or less prominent ridges separate the tectum, parietes, and intra-parietes from each other.

Upper Chalk, Norwich (common), *Mus.* Fitch. Northfleet (single spec.), Kent, *Mus.* J. Sowerby. Upper Chalk, Charing, Kent, *Mus.* Harris. Scania, and Quedlingburg in Westphalia, *Mus.* University, Copenhagen. Cyply bei Mons, Belgium, *Mus.* Brit. Gehrden Hanover, oberer Kreidemergel, *Mus.* Dunker and Roemer.

I have had far more difficulty in making up my mind regarding this the commonest cretaceous species, than with all the other fossil pedunculated cirripedes. From reasons previously stated, I have in this genus, when only separate valves have been found, taken the carina as typical. Comparing ordinary specimens of the carina of *Scalpellum maximum* and var. *sulcatum,* such as those figured in the 'Mineral Conchology,' I should certainly have considered them quite distinct, had not an examination of Mr. Fitch's fine collection from Norwich, together with several other specimens, shown me that there are intermediate forms which it is scarcely possible to class. Again, had I not seen a particular carina of *S. maximum var. cylindraceum,* in which the upper part displays a different character from the lower in the same individual valve, I should have unhesitatingly received it as a species, instead of, as I now do with certainty, as a mere variety. I feel, moreover, very great doubts whether the *S. lineatum* be a species, or merely another variety of *S. maximum;*

[1] For an explanation of this and all other terms, see the remarks on nomenclature and woodcuts in the Introduction, page 9 and 10.

its distinctive characters are extremely slight; but they do not blend away by any intermediate forms hitherto seen by me. Looking only thus far, it would have been natural to have classed, without any doubt, all the carinæ as varieties of *S. maximum*, but in the same Norwich beds, from which Mr. Fitch obtained his fine series of carinæ, there are scuta and terga, which undoubtedly belonged to the genus Scalpellum, and which, from being of equally large size, nearly equally numerous, and having a similar state of surface with the above carinæ, I believe belonged to them: but both the terga and scuta present a more remarkable range of variation than do even the carinæ. In the case of the terga, at one extreme of the series, I did not even at first recognise the valve to be a tergum! yet the forms so blend together with very short intervals, that I cannot specifically separate them. Terga of the two extreme forms come, also, from the same localities in Scania. In the case of the scuta there are three distinct forms in Mr. Fitch's collection, which I should certainly have considered as specifically distinct, had I not been led from studying the carinæ and terga to believe that this species varies much: moreover, the chief point of variation in the scuta, namely, in the character of the under surface of the upper part, I conceive to be, in some degree, in connection with one chief peculiarity in the terga, namely, the varying prominence of their occludent margins. Although I have not seen any other instance of so much variation in the scuta; yet I believe that I have taken the most prudent and correct course in describing them as mere varieties. From the more frequent coincidence of the carina, described as that of the true *maximum*, with the Varieties I of the scuta and terga, I believe that these valves belonged to the same individuals: with respect to the two other varieties, I have hardly any grounds for conjecturing which belonged to which. It is most unfortunate that not a single specimen of this species seems, hitherto, to have been found with all, or even a few, of its valves embedded together.

In giving names to the varieties, as judged of by the Carinæ, there is a difficulty in nomenclature; for the carina of *S. maximum* and of *S. maximum*, var. *sulcatum*, are apparently almost equally numerous in the Norwich beds; and might either be taken as typical of the species; I have chosen the former name, simply as having been more commonly used, and from this form having been apparently most widely distributed. I have described under it the original carina of *Pollicipes maximus* of J. Sowerby, and all the other valves, which I have reason to suppose belonged to this species. The other carinæ, however, as being in this genus the typical valve, are described under separate subordinate headings; the description of *S. maximum*, var. *sulcatum*, being given from Mr. Sowerby's original specimen. Under the typical *S. maximum*, I indicate as far as able, to which carinæ the varieties of the scuta and terga, there described, probably belonged.

Scalpellum maximum, *var. typicum.* Tab. II, figs. 1, 4, 5, 8.

S. carinâ introrsùm leviter arcuatâ, latitudine valvæ altitudinem superante; tecto transversè leniter arcuato; parietibus intra-parietibusque angustis, superficie pænè lævi.

Carina slightly bowed inwards; width of valve greater than the depth; tectum flatly arched transversely; parietes and intra-parietes narrow; surface nearly smooth.

Carina, Tab. II, fig. 1. In this, the typical variety, the carina is very slightly bowed inwards, widening gradually downwards from the apex, of which no part projected freely; walls rather thin; tectum very flatly arched, not sub-carinated; basal margin rectangularly or rather more acutely pointed; parietes very slightly concave, splaying outwards, narrower than one side of the tectum, separated from it and from the intra-parietes by rounded ridges; intra-parietes narrow, not extending baseward so far as the basal margin of the parietes; width of valve measured from marginal edge to edge, considerably greater than the depth, measured in the same place from the central crest to either marginal edge; but the width compared with the depth varies a little: inner margin of valve nearly straight. Length of longest specimen (*Mus.* Fitch) nearly $1\frac{1}{2}$ inch. This variety in the Norwich beds seems about equally common with var. *sulcatum,* but the former alone is found in Hanover and in Scania, excepting that in the latter country some specimens indicate a passage to the var. *cylindraceum.*

Scutum, Tab. II, fig. 8. In Mr. Fitch's collection there are three left-hand valves of a Scalpellum, which, from their size and smoothness, I have no doubt belonged to this species, and from their thinness, probably to the variety of carina considered as typical under the simple name of *S. maximum:* valve unusually thin and little convex; trapezoidal, with the apex less produced than is usual in the genus; broad in proportion to its length, the basal margin being ·66, and the occludent margin ·98 in length; therefore the breadth equals two thirds of the length. Basal margin (just perceptibly hollowed out) forming less than a right angle with the (just perceptibly outwardly arched) occludent margin, and forming an almost exact rectangle with the lateral margin; the latter meets the tergal margin at an angle of about 135°. The edge of the tergal margin is thickened and slightly reflexed; the upper part of the lateral margin is in some specimens a little bowed inwards. The baso-lateral angle is rounded and just perceptibly protuberant; no ridge or furrow runs from it to the apex. Internally the depression for the adductor muscle is singularly shallow (fig. 8 *c*); a very small portion of the upper part of the valve projected freely; the internal surface of the valve, above the pit for the adductor muscle, has not been thickened, and is therefore slightly concave or almost flat. The internal occludent edge in the upper part is only a very little widened, and is flat; on the tergal margin, a narrow ledge of about equal width with the occludent edge, marked likewise with lines of growth, must have overlapped the tergum. Largest specimen 1·15 in length.

Scutum, Var. II, Tab. II, fig. 9. This valve is narrow, moderately convex, with the upper portion much acuminated; the tergal margin is somewhat hollowed out, and is bordered by a narrow smooth slip, (as in the scutum of *S. arcuatum*,) which is simply formed by the thickening from within of the upper part of the valve; this slip does not reach to the uppermost point. The occludent margin is somewhat arched, at nearly right angles to the basal margin; lateral margin forming an angle a very little above a right angle with the basal margin. A conspicuous, curved, angular ridge runs from the apex to the baso-lateral angle, (which is not at all protuberant,) and divides the valve obliquely into two almost equal halves. Surface just perceptibly striated, finely and longitudinally. Internally there is a deep pit for the adductor scutorum, which is situate low down in the valve; the inner occludent edge in the upper part of the valve (*b*), above the adductor scutorum, widens suddenly, and is formed into a furrow, which, however, I do not believe to have had any functional importance; the central internal surface of the valve, above the pit for the adductor muscle, is somewhat prominent; and a quite small, almost flat, portion of the tergal side is marked by lines of growth, showing where it overlapped the tergum. Altogether there is a considerable resemblance between this valve, both externally, and more especially internally, and that of the *Pollicipes Angelini*. From the valve being acuminated, with the upper part rather solid, and from the surface being just perceptibly striated, it more probably belonged to var. *sulcatum* than to the typical *S. maximum*.

Scutum, Var. III, Tab. II, fig. 10. This third variety, of which the specimen is a fine large one, is about intermediate in outline or acumination between the first and second varieties: the tergal margin is thickened and reflexed as in the first, and is not bordered by a smooth narrow slip as in the second variety. There is no distinct angular ridge, as in the second variety, running from the apex to the baso-lateral angle. Internally the differences are more conspicuous; the depression for the adductor muscle is pretty well developed; a large portion of the upper part of the valve projected freely; the internal occludent edge, above the adductor-depression, becomes greatly widened and deeply hollowed out, but yet the furrow I believe, as in Var. II, to be of little or no functional importance, and merely a consequence of the internal thickening of the central upper part of the valve; on the tergal side a wide ledge shows the extent to which that margin overlapped the tergum. The internal surface of the valve, above the adductor-depression, is filled up solid and is exceedingly prominent, as is the ridge extending from it to the apex; this ridge, from the unusual width of the internal occludent edge, is pushed over to the tergal side of the valve.

Professor Steenstrup has sent me two small scuta, collected by M. Angelin at Kopinge and Balsberg, in Scania, which come near to the Third variety; the internal furrow, however, along the occludent margin, is much narrower, deeper, and oblique, so that it is partly covered by a lateral projection of the central portion: a tolerably distinct ridge runs from the apex to the baso-lateral angle. Amongst the several specimens from Hanover sent me by Drs. Dunker and Roemer, the scuta all belong to the First variety.

I believe all these differences in the scuta of the three varieties ensue partly from the varying acumination of the upper part, and consequently of the extent to which the apex projected freely, but chiefly from the degree to which the upper part of the valve above the adductor muscle has been internally thickened. In the first variety the upper part is simply concave, and the pit for the adductor very shallow; in the third variety, the same upper part is highly prominent, and apparently as a consequence the internal occludent edge is deeply furrowed; the pit for the adductor muscle is deepest in the second variety.

The above differences would perhaps affect the outline of the terga, but I am not able to follow the precise manner; nor should I have thought them sufficient to have produced the amount of variation presently to be described in the terga; but possibly other scuta may vary still more. At first I concluded that the upper part of the inner occludent edge, which in Var. III is deeply furrowed, received in it the occludent edge of the tergum (as the furrow on the *tergal side* of the apex of the scutum receives the edge of the tergum in the recent *Pollicipes mitella*), but this on consideration I do not think can possibly be the case, although it would amply account for the variation in the terga.

Terga. I have seen great numbers of these valves; eight specimens are in Mr. Fitch's collection from Norwich; one is figured by Mr. J. Sowerby in the 'Min. Conch.,' (Pl. 606, fig. 6,) and they are numerous in the collection from Scania and Hanover. These valves, which, as stated in the preliminary remarks, present a most remarkable amount of variation, will be best described under three distinct heads.

Variety I. Tab. II, fig. 5. This valve, from its greater width and smoothness, compared with the other varieties, perhaps belongs to the typical *S. maximum.* Surface smooth, with a mere trace of some longitudinal striæ, sub-rhomboidal, elongated, with the apex much produced and curled towards the carina; nearly flat; the occludent margin arched, nearly equal in length to the scutal margin; upper carinal margin hollowed out, about half the length of the lower carinal margin; the occludent and upper carinal margins meet each other at a very small angle, making the apex almost horn-like; from it to the bluntly pointed basal angle, a slight rounded ridge, and on the carinal side of it a slight furrow, (both becoming less plain towards the lower part of the valve,) extends. As seen internally, the thickness of the valve, in its upper part, varies; a rather large upper part projects freely. A rim along the occludent margin is rounded and slightly protuberant, with a slight depression in the valve parallel to it. Length of the largest specimen 1·2 of an inch.

This variety is found commonly near Norwich, in Scania, and Hanover.

Tergum. Variety II. Tab. II, fig. 6. The valve in this variety (from near Norwich) is much elongated, sub-triangular, approaching to crescent-shaped; lines of growth conspicuous, with a few very faint longitudinal striæ. Carinal margin not (or barely) distinguishable into an upper and lower portion; the whole being nearly straight, or very slightly concave. Apex extremely produced, narrow, and horn-like; curled towards the carina; apparently (for the apex is broken) a considerable portion was thickened, and must have projected freely. Occludent margin slightly arched, about equal in length to the scutal margin, which latter in the lowest part is curved and projects a little. Basal angle bluntly pointed. A rounded ridge (with a mere trace of a furrow on its carinal side), almost disappearing in the lower part of the valve, runs from the apex to the basal angle in a slightly curved course, strictly parallel to the carinal margin. The rim of the valve along the occludent margin is rounded and strongly protuberant, and, parallel to it, the surface is considerably depressed. Length of valve, when perfect, 1·2 of an inch. This variety differs from the first in the much greater straightness of the carinal margin, in the occludent rim being more protuberant, and in the scutal margin not being quite straight. One specimen presented a decidedly intermediate form, though rather nearer to the first than to the second variety.

Tergum. Var. III. Tab. II, fig. 7. The valves of this variety, of which I have seen five specimens, were for a long time quite unintelligible to me, and I at first even thought that perhaps they were rostral latera, but I now find that in outline, though not in general appearance, owing to their great thickness, they closely resemble the terga of *S. magnum.* One of the four specimens is almost exactly intermediate between the variety last named and that now to be described; hence there can be no doubt that they are really terga. The chief characteristic of the valves of this variety is their narrowness, and the solidity of their upper ends, which, together with a point of structure presently to be mentioned, makes me think it likely that they belonged to the individuals which possessed a carina, hereafter to be described under the name of *S. maximum,* var. *cylindraceum.* Valve smooth but with the lines of growth plain, extremely narrow, almost crescent-shaped; carinal margin considerably more concave than in Var. II, with a barely perceptible prominence in the upper part, marking the commencement of the freely projecting portion, and probably the point of upward extension of the carina. The occludent margin is arched, and is equal in length to the straight scutal margin. From the apex there runs a fine furrow (instead of a ridge and furrow, as in Vars. I and II,) to the basal angle, nearly parallel to the carinal margin, but almost blending with it in the lower part of the valve. The upper freely projecting portion is much thickened, and rendered almost horn-like, but to a variable extent; owing to this the width of the valve in the upper part also varies. In the specimens most characteristic of the present variety, the rim of the valve along the occludent margin is not at all, or barely, protuberant, nor is there any plain depression parallel to the occludent margin: in the intermediate specimen, however, above alluded to, the rim is protuberant and there is a plain depression, though both much less conspicuous than in the tergum of Var. II. On the internal surface of the upper freely projecting part, (marked with lines of growth,) there can be observed in two specimens a slight and variable longitudinal depression; judging from what occurs in the recent genus

Lithotrya, and from what may be faintly seen in the tertiary *Pollicipes carinatus*, and even in some specimens of *P. mitella*, I believe that this structure indicates that the upper freely projecting portion of the carina had its inside filled up and rendered prominent, which we shall see is the case with the carina of the variety *cylindraceum*. Length of largest specimen, eight tenths of an inch.

This variety is found at Norwich, in Scania, and at Cyply bei Mons, in Belgium.

Amongst the Scanian specimens from Kopinge (where the carina of the true *S. maximum* is commonly found) there are some terga differing from the variety just described, only in having the lower part of the valve less produced; and more especially in having on the internal surface of the upper part a smooth prominent ridge, lying rather nearer to the occludent than to the carinal margin, and therefore in exactly the same position in which a little group of small, sharp, longitudinal ridges occurs in the terga of *S. arcuatum* and of some other species. I am surprised at such a point being variable, but I cannot doubt that this valve belongs to the same species. I may add that it was this trifling point of structure, which first led me to suspect that these singular crescent-shaped valves were really terga. Finally, I may remark, that when all the ten terga now described are placed in a row, it is scarcely possible to doubt that they form merely varieties of the same species.

Carinal latus, Tab. II, fig. 4. This is the only valve which remains to be described, for neither the rostrum nor rostral latera are as yet known. It was found at Kopinge, in Scania, where the carina of the true *S. maximum* occurs abundantly; it was sent to me by Professor Steenstrup, who attributed it to this species. I have also seen a specimen from Hanover, where the carina of the true *S. maximum* is also found, and another small specimen from Charing, in Kent. Valve thin, of an irregular shape, sub-triangular; flat, except at the umbo, which projects outwards, owing to a ledge formed beneath and round it; carinal margin very slightly convex, with a linear furrow parallel to it, between which and the edge the lines of growth are abruptly upturned; lower margin considerably convex; upper margin slightly concave, with a slight depression parallel to it, between which and the edge the lines of growth are rectangularly reflexed towards the umbo. The two Scanian specimens differed slightly in outline; chiefly with respect to the projection of the ledge round the umbo. Width of largest specimen one quarter of an inch. This valve unmistakeably resembles the homologous valves in *S. quadratum* and *fossula*, but can be distinguished from both; the end opposite the umbo is much less produced than in *S. quadratum;* the whole valve is wider, and the furrows much less developed, than in *S. fossula*, to which it comes nearest.

Affinities. Before describing the several varieties as characterised by their carinæ, I will offer a few remarks on the affinities of this, the most common and widely distributed species of all the cretaceous pedunculated cirripedes. Mr. James Sowerby at first naturally described it as a Pollicipes; quite lately in Mr. Dixon's work he has considered it as belonging to the same genus with his eocene *Xiphidium quadratum* and our *Scalpellum quadratum*. Still closer is the affinity with the cretaceous *S. fossula;* the carinæ of both have intra-parietes; the tectum is distinct from the parietes, which latter are either channelled or concave; the trapeziform scuta of *S. quadratum, fossula,* and *maximum,* are unmistakeably alike, and even more striking is the resemblance of the carinal latera; there can be no doubt of these three species belonging to the same genus, and having the same number of valves, namely, as I have shown under *S. quadratum* and *fossula,* probably twelve.

Geological History. This species, with its varieties *cylindraceum* and *sulcatum,* is very common in the Upper Chalk strata of Norwich; I have seen one specimen from the Upper Chalk of Northfleet, in Kent. It is common in the sandstone beds of Scania, which I am assured by Professor Forchhammer, are without doubt equivalent with the Faxoe beds, and therefore belonging to a stage above our flinty chalk. I have seen, also, one specimen, belonging, I believe, to this species, from the same stage in Westphalia; and another from Belgium; it is also common at Gehrden, in Hanover, in the 'Oberer Kreidemergel' of Roemer.

Scalpellum maximum, var. cylindraceum. Tab. II, fig. 2.

S. parte superiore carinæ liberè prominente, parte interiore intra-parietibus rotundatis, inflexis, itá repletá, ut pæne cylindrica fiat; superficie externá lævi, tecto parietibusque pæne confluentibus.

Carina, with the upper portion projecting freely, and with the inside filled up by the rounded inflected intra-parietes, so as to be almost cylindrical; exterior surface smooth, with the tectum and parietes almost confluent.

Amongst the specimens from Norwich, two differed from the others in being a little more elongated and smoother, in the parietes becoming almost confluent, low down on the valve, with the tectum, and in the intra-parietes being very little developed. On the internal face this variety presents its most remarkable character; for a large upper portion of the valve must have projected freely, and the intra-parietes, instead of forming a thin wall on each side, are thickened, rounded, and turned inwards, so as almost to meet, and thus to fill up the original concavity of the valve. Hence a section (fig. 2, *c*) of the upper part, some way below the apex, is almost cylindrical, or more strictly oval with the longer axis in the longitudinal plane of the animal, with either a wedge-formed hollow, or a mere, almost closed, cleft on the under side, penetrating not quite to the centre of the solid valve. The two specimens differ, one in being in a transverse line exteriorly much depressed, the other highly arched or convex, and internally still more conspicuously in the degree to which the intra-parietes have filled up the upper part. In one of the specimens there is even a difference on the opposite sides of the same individual valve. Notwithstanding these varieties, I should have much hesitated to have ranked these peculiar carinæ under *S. maximum*, had not the upper part in one specimen actually retained all the usual characters of *S. maximum*, the precise line where the manner of growth had changed, being distinctly visible. It is represented in Plate II, fig. 2, *a* and *b*. Amongst the Scanian specimens, some make an approach to this variety.

SCALPELLUM MAXIMUM, VAR. SULCATUM. Tab. II, fig. 3.

POLLICIPES SULCATUS. *J. Sowerby.* Min. Conch., pl. 606, solummodo, fig. 2. Fig. 7 fortasse Carina *P. Angelini.* Fig. 1, Tergum fortasse *P. striati.*[1]

S. carinâ introrsùm valde arcuatâ, sub-carinatâ; valvæ latitudine circâ dimidium altitudinis æquante, tecto transversè præruptè arcuato; parietibus intra-parietibusque latiusculis. Apice solidè repleto, liberè paululùm prominente; superficie externâ striis paucis, rotundatis, ad alterum vel utrumque latus costarum duarum tectum et parietes separantium.

Carina considerably bowed inwards, subcarinated; width of valve about half of the depth; tectum in a tranverse line, steeply arched; parietes and intra-parietes rather wide; apex filled up solid, and projecting freely a little; exterior surface with a few rounded striæ on either one or both sides of the two ridges which separate the tectum and parietes.

Having had the advantage of seeing Mr. J. Sowerby's original specimen, the valve now to be described is certainly that figured by him as *Pollicipes sulcatus.* As already stated, certain specimens of this variety differ strikingly from the carinæ typical of *S. maximum;* whereas others, from the same formation and locality, are so intermediate that they can, with difficulty, be arranged on either side: this is also the case with one from Cyply bei Mons, in Belgium. This variety is common in the Upper Chalk of Norwich.

In a well-marked specimen of this variety, the chief distinctive characters, as contrasted with the true *S. maximum*, consist in the tectum being more steeply arched, in the depth of the valve being much greater than the width, in the intra-parietes and parietes being more developed, in the whole valve being more bowed inwards, in the walls being thicker and apex filled up solid, in the surface having a few fine raised lines on each side of the ridge separating the tectum and parietes, and, lastly, in the tectum being sub-carinated.

[1] If I am correct in considering the carina of *P. sulcatus* to be only a variety of that of *S. maximum*, the tergum figured by Mr. Sowerby as belonging to his *P. sulcatus* cannot so belong; for it does not at all resemble the homologous valve of *S. maximum.* I believe from the character of the ridge running from the apex to the basal angle, that it belonged to a Pollicipes, which must have been coarsely striated longitudinally, and therefore I have provisionally described it under *Pollicipes striatus.*

Carina moderately bowed inwards, widening gradually downwards from the apex, of which a small portion is filled up solid, and must have projected freely; walls moderately thick; the two sides of the tectum are rather steeply inclined to each other, and meet in a central line, which is subcarinated with a slightly prominent ridge; basal margin rectangularly pointed; parietes nearly flat, about as wide as the tecta, in some specimens perpendicular, so as not to be visible when the valve is viewed from a central dorsal point; in others, very steeply splayed outwards; separated from the intra-parietes by a conspicuous rounded ridge, and from the tectum by a nearly equally large ridge, which has generally one, two, or three fine, longitudinal, raised lines on either one or both sides of it: in one specimen the whole surface was thus coarsely and obscurely lined. The intra-parietes are rather wide, extending to the basal margin of the parietes. Depth of valve, measured from the central crest to either inner edge, is about equal to the entire width, as measured from inner edge to edge. The depth compared with the width, though the most conspicuous character, varies a little. Inner edge of valve nearly straight. Length of longest specimen (in *Mus.* Bowerbank) 1·6 of an inch. This is the largest carina I have seen in any fossil cirripede.

5. SCALPELLUM LINEATUM. Tab. II, figs. 11 and 12.

S. superficie totâ carinæ lineis tenuibus, rotundatis, longitudinalibus, proximis, microcoscopicis obtectâ; cristæ centralis costâ crassiore; costis duabus vel tribus tectum et parietes separantibus; latitudine valvæ circa dimidium altitudinis æquante; intra-parietibus latiusculis, nullâ costâ conspicuâ a parietibus separatis; apice solidè repleto, aliquantulum liberè prominente.

Carina with the whole exterior surface covered with fine, rounded, longitudinal lines, scarcely visible to the naked eye; with a thicker ridge on the central crest, and with two or three similar ones separating the tectum and parietes; width of valve about half of depth; subcarinated; inter-parietes rather wide, not separated by a conspicuous ridge from the parietes. Apex filled up, solid, and projecting freely a little.

Lower Chalk of Sussex, *Mus.* J. Morris; *Mus.* J. Sowerby.

I have seen two carinæ in the collections of Mr. Morris and Mr. J. Sowerby so exactly like each other, and having a somewhat different aspect from *S. maximum*, var. *sulcatum*, to which they come nearest, that they deserve to be described, whether or not they are really specifically distinct. I long hesitated whether to give them a specific name, and have been, in some degree, influenced in doing so, from the presence of scuta and terga in the Lower Chalk, which indicate a distinct but closely-allied species. The scutum is in Mr. Morris's collection, and came in the same lot with the carina from Sussex: the tergum is in Mr. Bowerbank's collection from the Lower Chalk of Maidstone. These valves are marked with longitudinal raised striæ more plainly than is the carina.

Carina (fig. 12); moderately bowed inwards; inner margin nearly straight; widening very gradually downwards from the apex, of which a very small part is filled up solid, and must have projected freely; walls rather thin. Both tecta and parietes are regularly striated longitudinally, with raised, hair-like, fine lines scarcely visible to the naked eye; one central, and two or three on each side between the tectum and parietes, being about twice as large as the others, and visible to the naked eye. Tecta rather steeply inclined towards each other; central line sub-carinated; basal margin rectangularly pointed; parietes slightly concave, about as wide as half the tectum; steeply inclined outwards; separated from the intra-parietes on each side by a slight ridge. Intra-parietes set a little inwards, wider in the widest part than the adjoining parietes or tecta; extending baseward not as far as the basal margin of the parietes. Depth of valve measured from central crest to either inner edge, nearly equal to the entire width, as measured across from inner edge to edge. In many respects this carina is intermediate between those described under *S. maximum* and *S. maximum*, var. *sulcatum;* but comes nearest to the latter: the intra-parietes not extending so far baseward; and the delicately lineated exterior surface gives it, however, a somewhat different aspect.

Scutum (fig. 11); this valve, from the Lower Chalk of Sussex, resembles that of *S. arcuatum;* its surface is covered with raised striæ, which are further apart, and less plain than in the typical specimens of *S. arcuatum* from the Gault, but resemble those in the variety from the Grey Chalk of Dover. Outline trapezoidal: the baso-lateral angle is very broad, rounded, and protuberant; no ridge runs from it to the apex: the basal margin projects very slightly close to the rostral angle, and the tergal margin is not inflected as in *S. arcuatum.* The internal surface of the valve, along the tergal margin, is not furrowed or marked by lines of growth: I have no doubt that this valve is, at least, distinct from *S. arcuatum.*

Tergum. This valve, from the Lower Chalk of Maidstone, resembles that of *S. arcuatum,* var. from the Grey Chalk; it is, however, slightly more elongated: it further closely resembles a tergum, which I have provisionally attributed to Pollicipes striatus, differing from it in being less elongated, and more especially in the absence of a ridge, steep on the carinal side, which in that species runs from the apex to the basal angle.

Finally, I may remark, that these three valves, on the supposition that they have been rightly attributed to one species, indicate a form intermediate between *Scalpellum maximum* of the Upper Chalk, and *S. arcuatum* of the Grey Chalk and Gault.

6. Scalpellum hastatum. Tab. II, fig. 13.

S. carinâ intra-parietibus, intrà paululùm positis, instructâ; valvâ totâ introrsùm valdè arcuatâ, margine interno non recto; margine basali acuto, lanceolato; valvâ tenui, lævi, tecto transversè leniter arcuato; parietibus à tecto vix disjunctis.

Carina having intra-parietes set a little inwards; whole valve much bowed inwards, with the inner margin not straight; basal margin sharply pointed, spear-shaped; valve thin, smooth; tectum in a transverse line flatly arched; parietes barely separated from tectum.

Grey Chalk, Dover, *Mus.* Brit.

Carina smooth, narrow, furnished with intra-parietes, widening gradually from the apex downwards; extremely much arcuated, so that the uppermost part, of which none, or very little, projected freely, is at right angles to the basal part. Not only is the dorsal surface considerably arcuated, but so are the inner margins, which is much more important. Basal margin sharply pointed, with the two edges meeting each other at about an angle of 75°. Roof with the two sides continuously and flatly arched; parietes rather narrow, slightly concave, barely separated from the tecta: the concavity of the parietes, as seen on the basal margin, together with the sharpness of the central portion, makes the lower part of the valve spear-shaped. The intra-parietes are set a little within the parietes; they extend down to about two thirds of the entire length of the carina, and not to the basal margin of the parietes: they are widest at about only one fourth of the entire length of the carina from the apex, and here they equal in width the rest of the valve. Internally the valve is, in the upper part, owing to the wide intra-parietes, deeply concave; in the lower part, only slightly so.

Length of carina, measured along the chord of the arch, ·75 of an inch.

Affinities. This species certainly comes very near to *S. maximum;* but I think it is distinct, and is its representative in the Grey Chalk. I have seen only a single specimen. The carina differs from the former varieties and species in its smoothness, thinness, in the acumination of the basal margin, in its much arcuated form, and more especially (for this, probably, would greatly influence the outline of the terga,) in the inner margins being also thus arcuated.

7. Scalpellum angustum. Tab. I, fig. 2.

Xiphidium angustum. *Dixon.* Geology of Suffolk, tab. xxviii, fig. 9.

S. carinâ angustâ, introrsùm valdè arcuatâ; tecto à parietibus rectangulè inflexis costâ, (ut videtur) disjuncto; intra-parietibus usque ad dimidium valvæ pertinentibus, deinde obliquè et abruptè truncatis; margine basali acutè cuspidato.

Carina narrow, much bowed inwards; tectum apparently separated from the rectangularly-inflected parietes by a ridge; the intra-parietes extend down half the valve, and are there obliquely and abruptly truncated; basal margin sharply pointed.

Chalk.

I know this species only from the plate in Mr. Dixon's work. Being well aware of Mr. J. de C. Sowerby's great accuracy, I cannot doubt that the intra-parietes are at their lower end, abruptly and obliquely truncated in the manner represented in the Plate: this character, with its sharply-pointed basal margin, makes me believe the species to be new: it comes, I imagine, nearest to *S. hastatum.*[1]

9. SCALPELLUM TRILINEATUM. Tab. I, fig. 5.

S. carinæ tecto transversè leniter arcuato, subcarinato, costâ centrali et costis duabus lateralibus, rotundatis, tumidis; parietibus angustis leviter concavis, rectangulè inflexis.

Carina, with its tectum in a transverse line flatly arched, sub-carinated, with a central and two lateral, rounded protuberant ridges; parietes narrow, slightly concave, rectangularly inflected.

Grey Chalk, Dover, *Mus.* Brit., Flower. Chalk Detritus, Charing, Kent, *Mus.* Harris.

Carina (fig. 5, *a*—*d*); moderately arched, narrow, gradually widening from the apex to the base, plainly marked by lines of growth: no part apparently projected freely. The tectum is flatly arched, sub-carinated, with its central crest forming a rounded protuberant ridge; on each side, the tectum is bounded by similar, very slightly larger ridges, making

[1] 8. SCALPELLUM QUADRICARINATUM.

POLLICIPES QUADRICARINATUS, *Reuss*, Verstein. Bohmisch. Kreideformation (1846), Tab. xlii, fig. 18.

S. carinâ intra-parietibus latis (ut videtur) instructâ; tecto transversè plano, lævi, costâ prominente utrinque marginato; margine basali abruptè truncato.

Carina having apparently wide intra-parietes; tectum in a transverse line, flat and smooth, bordered on each side by a prominent ridge; basal margin abruptly truncated.

Bohemia. Untern Plänerkalke (Chalk-marl).

I know this species only from an imperfect plate, but good description of a carina, in Reuss' work: it is an interesting form, showing in its truncated basal margin and flat tectum a still closer affinity to the recent *S. rutilum,* even than does *S. fossula;* thus confirming the view I have taken of the affinities of these several species.

Carina; rather narrow, slightly bowed inwards: tectum quite flat and smooth, separated from the parietes by a smooth prominent ridge: parietes concave, rectangularly inflected: intra-parietes apparently well developed, separated from the parietes by a ridge: basal margin abruptly truncated.

together three ridges. The basal margin is bluntly pointed, with the two sides meeting each other at an angle of rather above 90°. Parietes rather narrow, rectangularly inflected, slightly concave: in the upper part there is no trace of intra-parietes.

Terga (fig. 5, *e—i*). In Mr. Flower's collection there is a tergum, (embedded in exactly the same matrix,) which, from a certain degree of resemblance in outline with that of *S. arcuatum*, the nearest congener to *S. trilineatum*, and from another point of resemblance with *S. fossula*, I believe belonged to this species.[1] The valve is very smooth, with obscure traces of fine striæ radiating from the umbo; nearly flat; pointed oval, but with the scutal side much more protuberant than the carinal. Apex much acuminated, curled forwards; carinal margin much and regularly bowed from the upper to the basal point, which latter is blunt and square: from it to the apex there runs, in a curved line, nearly parallel to the carinal margin, a barely perceptible broad ridge. Occludent margin curved up towards the umbo, short compared to the scutal margin; parallel to it there runs a very wide and very shallow depression. Scutal margin, with a portion corresponding with the above depression, forming rather more than a third of the margin, not projecting so much as the lower two thirds, and separated from this lower part by a slight bend, probably marking the spot to which the apex of the upper latera extended.

Affinities. The carina obviously most resembles that of *S. fossula* and *arcuatum;* it differs plainly from both, in having a central rounded ridge: in the two well-developed boundary ridges of the tectum it comes nearest to the cretaceous *S. fossula;* but in the absence of the intra-parietes (and this I conceive is a more important character), it comes nearest to the *S. arcuatum*, from which, however, it can be at once distinguished by the absence of the longitudinal striæ. The tergum above described, which I believe belonged to this species, in the form of the scutal margin, comes nearest to that of *S. fossula*, though in general shape perhaps nearer to *S. arcuatum*. In *S. fossula* the carina has intra-parietes, which are closely adjusted to the straight carinal margins of the terga: in *S. trilineatum* the intra-parietes are absent, but in their place the carinal margins of the two terga are themselves highly protuberant, so that in these two species, although the upper parts of the carinæ and terga are separately of very different shapes, they give, when combined together, a similar general outline.

10. Scalpellum simplex. Tab. I, fig. 9.

S. carinâ lævi; parietibus angustissimis, rectangulé inflexis; tecto subcarinato, transversè mediocriter arcuato; margine basali rectangulè acuminato.

[1] It must however be added that the terga, at present unknown, of *S. hastatum*, a species occurring in the Grey Chalk of Dover, would probably have the same outline, and almost certainly would have a very smooth surface.

Carina smooth; parietes extremely narrow, rectangularly inflected; tectum sub-carinated, in a transverse line moderately arched; basal margin rectangularly pointed.

Lower Greensand, Maidstone. Mus. J. Morris.

I know this species only from a single carina, which is chiefly characterised by its simplicity: it is, I think, certainly distinct from all the others. In the sides of the carina being simple, that is in not being divided by a ridge into parietes and intra-parietes, it comes nearest to *S. arcuatum* and *trilineatum*, from the former of which it is readily distinguished by its smoothness; and from *S. trilineatum* by the absence of the three ridges. This species possesses some interest, as being the oldest cirripede, which I have ventured to attribute to the genus Scalpellum. *Carina* moderately tapering, slightly bowed towards the terga; sub-carinated, but with the central ridge smooth; transversely moderately arched; basal margin rectangularly pointed; the whole surface is smooth. Parietes extremely narrow, rectangularly inflected, set inwards, not extending down to the basal margin, with the lines of growth almost parallel to the inner edges of the valve.

11. Scalpellum arcuatum. Tab. I, fig. 7.

S. valvarum lineis angustis elevatis ab apice radiantibus: carinæ tecto transversè leniter arcuato, et parietibus rectangulè inflexis, leniter concavis, lævibus.

Valves with narrow elevated lines radiating from their apices. *Carina* with its tectum in a transverse line flatly arched, and with the parietes rectangularly inflected, slightly concave, smooth.

Gault, Folkstone, *Mus.* Bowerbank, J. Sowerby, Flower. Var. in Grey Chalk, Dover, *Mus.* Brit. Pläner (Chalk-marl) near Hildesheim. Mus. Roemer.

I have ranked this species under Scalpellum instead of Pollicipes, from the somewhat greater resemblance of its scuta and carina with the fossil species of Scalpellum, than with any known Pollicipes; though in some respects it appears rather intermediate. This species appears to come nearest to the *Pollicipes radiatus* of J. de C. Sowerby in 'Geol. Trans.,' vol. iv, 2d Series, Pl. XI, fig. 6, but besides that that species comes from the Lower greensand, the lower angle of its tergum is much more pointed; the upper figure of the two there given appears to be something wholly different. From the state of the specimens, I believe that the three following valves, all in Mr. Bowerbank's collection, belonged to the same species.

General Appearance. Carina, scuta, and terga plainly marked with prominent, very narrow, straight ridges, radiating from their apices; the interspaces between these ridges are three or four times as wide as the ridges themselves; the lines of growth are very fine and narrow.

Carina (fig. 7, *a, b, g*); narrow, considerably arched: tectum flatly arched, obscurely subcarinated: parietes rectangularly inflected, somewhat concave, and *not longitudinally ridged,* like the tectum, about two thirds as wide as half the tectum: basal margin bluntly pointed, the two edges meeting each other at rather above a right-angle; a trace of a rounded ridge separates the tectum and parietes; in the upper part of the carina there is no trace of intra-parietes, therefore the section of the upper half of the carina is only four-sided, see fig. 7, *g*.

Scutum (fig. 7, *f*); moderately convex, with the apex acuminated: lateral margin nearly parallel to the slightly arched occludent margin, and at right angles to the straight basal margin; a distinct ridge runs from the apex to the baso-lateral angle, which is distinctly prominent and rather sharp. The valve, above a line running from the apex to the tergo-lateral angle, is inflected; and the narrow portion thus inflected, which cannot be seen when the valve is viewed from above, is destitute of the longitudinal ridges.

In a specimen from the Grey Chalk of Dover, in which the internal surface was visible, there was, above the well-marked depression for the adductor muscle, a prominent, central, slightly oblique ridge, with the inner occludent edge of the valve widened and slightly hollowed out on the one side, and with a trace of a furrow on the other or tergal side.

Terga (fig. 7, *c, d*); flat, oval, with the scutal angle rather protuberant; basal angle not sharply pointed, from it to the apex there runs an obscure furrow, which furrow in the lower part of the valve is central, but higher up is situated at about one third of the width of the widest part of the valve from the carinal margin; in the lower part of the valve, the lines of growth (and consequently the margins of the valve) make with this furrow, equal angles on its opposite sides. The valve is slightly depressed, parallel to the occludent margin. A small portion of the apex projected freely; *internally,* in the upper part, rather nearer to the occludent than to the carinal margin, the valve is prominent, and this part is marked with two or three little ridges (*c*) ending abruptly downwards.

Size of largest specimen,—length of carina, ·85 of an inch; of scutum, from the apex to the basal margin, rather above ·6; of terga, ·55. I do not, however, know that these valves belonged to the same individual.

Variety. In the British Museum there is a scutum, and in Mr. Flower's collection there is a tergum, both from the Grey Chalk of Dover, which are most closely allied to, if not identical with, the above valve. The raised striæ on both are rather further apart and are less prominent. In all the other characters the scutum is identical. The tergum differs in its carinal margin, being rather more angularly bent, and in there being no furrow running from the apex to the basal angle; but these differences are trifling and insufficient for distinguishing a species. Amongst some specimens most kindly sent me by Roemer, there is a tergum from the Pläner of Sarstedt (Chalk-marl), which is identical with this.

Affinities. This species is related to *S. trilineatum, simplex,* and *solidulum,* in the absence of intra-parietes; in the terga it comes closest to the latter species.[1]

[1] 12. SCALPELLUM SOLIDULUM. Tab. I, fig. 8.

POLLICIPES SOLIDULUS. *Steenstrup* in Kroyer's Tidsskrift, b. ii (1839), pl. v, fig. 14 et 14*.
— UNDULATUS. *Id.* Id. Id. fig. 6.

S. valvarum lineis latiusculis elevatis ab apiee radiantibus. Carinæ parte superiori liberè prominente, et cristâ centrali, internâ, longitudinali instructâ.

Valves with rather wide elevated lines radiating from their apices. Carina, with the upper part freely projecting, and internally urnished with a central prominent, longitudinal crest.

Scania (Kjuge). Mus. Univers. Copenhagen.

Professor Steenstrup has described under this name some carinæ, in so worn a condition, that I confess that I thought it quite impossible to characterise them; and under the name of *P. undulatus,* some well-preserved terga. Quite lately, M. Angelin has sent to Professor Steenstrup, from Kjuge in Scania, several of the same carinæ in a much better condition, a scutum, and some broken terga of *P. undulatus,* which, from the similarity of their longitudinally striated surfaces, M. Angelin believes belonged to the same species: I quite concur in the probability of this view. The better state of the carinæ proves the sagacity of Professor Steenstrup, in considering his worn specimens indicative of a distinct species. Had I seen these carinæ alone, I should have much hesitated in considering them as belonging to a Scalpellum: for they differ considerably from the same valve in all other species; the parietes, or rather the part answering to the parietes, being here so much inflected, that they fill up and render solid the upper part of the valve; but the scutum undoubtedly belonged to a Scalpellum, and the terga closely resemble the same valve in the *S. arcuatum.*

Carina (fig. 8, *b, c, d*); narrow, elongated, strong and solid; moderately bowed inwards; basal margin rectangularly pointed; surface covered with rather broad slight ribs, central one being apparently (for the best specimens are much worn) twice as broad as any of the others. In a transverse line, the tectum is considerably arched in the upper part of the valve, and only slightly arched in the lower part. A considerable length of the upper part must have projected freely; this portion being filled up solid, and having a central, largely prominent crest or ridge: it appears, for the specimens are in a much worn condition, as if the ridge had been formed by the inflection of the parietes on each side, and their perfect junction. In the peculiar and almost monstrous variety of *S. maximum,* called var. *cylindraceum,* we have nearly the same structure; a cleft, however, being left, marking the line of junction of the opposite parietes. In general appearance and proportions, this carina comes nearest to those of Scalpellum; but in the peculiar modification of the parietes (if they can be so called) into a central crest, and in the apparent (from worn state) absence of any distinct ridge separating the tectum and parietes, the valve departs from the general description of the carina in Scalpellum.

Scutum; of this valve, which undoubtedly belonged to a Scalpellum, there is one entire specimen, but with the angles so much rounded, that I can point out no distinguishing character from the same valve in *S. arcuatum* (fig. 7, *g*), of which a figure has been given, except that the longitudinal ridges are proportionally broader and further apart. The ridges closely resemble those on the above-described carina.

Terga (fig. 8, *a*); sub-triangular, flat, strong, and thick, with moderately wide, not quite straight ridges, radiating from the apex: the interspaces between the ridges are three or four times as wide as the ridges themselves; valve very slightly depressed, parallel to the occludent margin. A slight ridge, connecting the sharp basal apex, runs quite close to the carinal margin, even in the lower part of the valve: in

13. Scalpellum tuberculatum. Tab. I, fig. 10.

S. valvarum lineis tenuibus, tuberculatis, elevatis, ab apice radiantibus: carinæ tecto transversè leniter arcuato, et parietibus striatis: scuti umbone prope in medio marginis occludentis posito, costis duobus ab umbone ad angulum basi-lateralem, et ad basalis marginis medium decurrentibus.

Valves, with fine, tuberculated, elevated lines, radiating from their apices: carina, with the roof in a transverse line, gently arched, and with the parietes striated: scutum, with the umbo placed nearly in the middle of the occludent margin, with two ridges running from the umbo to the baso-lateral angle and to the middle of the basal margin.

Chalk Detritus. Charing, Kent. *Mus.* Harris.

Through the kindness of Mr. Harris, I have examined several valves, which I believe to belong to the same species: the specimens were found in the chalk detritus, and, therefore, may have come from the Upper or Lower Chalk or Chalk-marl; but more probably from the Upper Chalk. With respect to the scuta and terga I have scarcely any doubt, from certain peculiarities, that they belonged to the same species; but with regard to the most important valve, the carina, I cannot feel quite so certain: when the latter is so held, that the parietes are not visible, it has a very close general resemblance to the same valve in *Pollicipes rigidus*. In the carina, the present species comes closer to *S. arcuatum* than to any other species; in the other valves, especially in the singular scuta, it departs widely from that and all other known fossil forms, with the exception of *S.* (?) *cretæ*, of Denmark. All the specimens which I have seen are small; the carina being ·2 long, and the terga less than ·15 of an inch in length, in the largest specimens.

consequence of this, the lines of growth make a different angle, on the opposite sides, with this ridge: as the valve has been somewhat worn, it is possible that the carinal margin may have been more abraded than is apparent. Internally, it is seen that a considerable portion of the upper part of the valve projected freely; beneath this, the inner surface is slightly convex, but smooth, and though the shell has been much worn, I doubt whether there ever existed ridges, as on the internal surface of the upper part of the terga in *S. arcuatum*, to which valve this presents a close general resemblance. Length of tergum (when perfect), 1·2 of an inch.

Carinal Latus (fig. 8, *e, f*); amongst the fossils from Kopinge (at which place the same species are found as at Kjuge), there is a valve, which I believe to be a carinal latus of a Scalpellum, and which, from its longitudinal ridges, more probably belonged to the present than to any other species: from its peculiarity it is in any case worthy of description. In form it is a segment, somewhat less than a quarter, of a circle; of this segment, nearly half (I believe the upper half) has its end or circumferential margin much hollowed out, and its surface smooth: the other half has its periodical growth-ridges very prominent, and these are crossed by a few slight longitudinal ridges. One of the lateral sides (the upper, I believe,) is reflexed so as to form a prominent ledge; the other side is slightly inflected.

The *valves* all have their surfaces plainly ribbed longitudinally; the ribs are narrow, and as they cross each zone of periodical growth they are tuberculated.

Carina (fig. 10, *b*, *c*); narrow, tapering, little bowed inwards; tectum in a transverse line, steeply arched, not carinated; basal margin bluntly pointed; in very young specimens, however, it is evident from the lines of growth, that the basal margin must have been rounded; the parietes are inflected, and rather narrow, being barely half the width of half the tectum; they are plainly marked by parallel lines of growth; internally the valve is rather deeply concave; no part of the apex projected freely.

Scuta (fig. 10, *e*); umbo of growth on the occludent margin, at about one third of the entire length of the valve from the apex; somewhat convex; four-sided, the margins consisting of the lateral, which is considerably longer than the other sides; the basal which forms nearly a right angle with the lower half of the occludent margin; and of an upper and lower occludent margin, meeting each other at about an angle of 135°: the margin which I have here called the upper occludent, homologically corresponds with the tergal margin of the other cretaceous species, and with the upper, nearly straight, portion of the occludent margin in the tertiary *S. magnum* and the recent *S. vulgare*,—a fact which has been mentioned under the head of Scalpellum. The edge of the upper occludent margin forms a strongly prominent ridge, with its apex forming a slight projection; a second less prominent ridge runs from the umbo to the baso-lateral angle, and a third faint ridge runs from the umbo to a point in the basal margin, nearer to the rostral than to the baso-lateral angle. Internally there is a rather deep hollow for the adductor muscle; along the under surface of the upper occludent margin there is a slightly prominent ridge, bordered by two slight depressions.

Terga (fig. 10, *a*); flat, elongated diamond-shape; close and parallel to the occludent margin there is a narrow, very prominent ridge or plait, the end of which forms a slight projection; a straight ridge runs from the apex to the sharp basal angle; the scutal and lower carinal margins are of equal length, and longer than the occludent and upper carinal margins, which latter are equal, and meet at an angle very slightly less than a rectangle. On the under surface there is a slight depression and ridge, close and parallel to the occludent margin. I have no doubt that the ridge along the upper occludent margin of the scuta, and that on the occludent margin of the terga, together with their projecting points, are related to each other, owing to the close contact of these valves.[1]

[1] 14. Scalpellum semiporcatum. Tab. I, fig. 6.

S. carinâ ignotâ: scuti costis duobus ab umbone ad angulum basi-lateralem et ad marginis basalis medium decurrentibus: superficie inter hanc costam et marginem occludentem lineis tenuibus, longitudinalibus, elevatis instructâ.

Carina unknown: scutum, with two ridges running from the umbo to the baso-lateral angle, and to

15. Scalpellum (?) cretæ. Tab. I, fig. 11.

Anatifera cretæ. *Steenstrup.* Kroyer's Tidsskrift, 1837 et 1839, b. ii, pl. v, figs. 1, 2, 3.

S. valvis lævibus tenuissimis: scuti umbone propè medium marginis occludentis posito; costis tribus obscuris ab umbone ad angulos tergo-lateralem et basi-lateralem, et ad medium marginis basalis decurrentibus: carinæ apice et margine basali acutis; distincti parietes absunt.

Valves smooth, extremely thin: scutum with the umbo placed nearly in the middle of the occludent margin, with three obscure ridges running from the umbo to the tergo-lateral and baso-lateral angles, and to the middle of the basal margin: carina with the apex and basal margin sharply pointed; without distinct parietes.

White Chalk, Denmark, *Mus.* Univers., Copenhagen. Chalk Detritus, Charing, Kent (?), *Mus.* Harris.

Preliminary Remarks. I owe to the kindness of Professor Steenstrup, as in so many former instances, an examination of several specimens of this fossil, which is of interest, as being extremely common and characteristic of the white chalk of Denmark. Amongst the numerous minute specimens from the chalk detritus of Charing in Kent, sent me by Mr. Harris, there are some carinæ so similar that I have ventured, with doubt, to rank this as a British species; the carina, however, in this species, are far from characteristic. I have felt much hesitation in admitting this species in the genus Scalpellum: Professor Steenstrup was originally inclined to believe that the capitulum was formed of only five valves; could this be proved, the species would very naturally rank with a small recent one from the Island of Madeira, which, owing to the upward growth of the scuta, and to certain peculiarities in the animal's body, I have felt myself compelled to raise to the rank of a genus, under the name of *Oxynaspis.* But as the valves of *S.* (?) *cretæ* have never been found united, and as the main ones are very small, fragile, and generally in a broken condition, the small lower ones might easily be overlooked. I have seen, indeed, in two instances,

the middle of the basal margin; the surface between the latter ridge and the occludent margin covered with fine longitudinal elevated lines.

Scania (Kopinge). *Mus.* Univers., Copenhagen.

I have in this one instance departed from my rule of never naming any other valve, except the carina in the genus Scalpellum; but the scutum here to be described almost certainly belongs to this genus, and is interesting in connection with the homologous valves in *S. tuberculatum* and *S.* (?) *cretæ,* to which species it is apparently allied, but yet differs greatly from them in the umbo being seated at the uppermost point of the valve.

Scutum, moderately elongated, slightly convex; a narrow, prominent, well-defined ridge runs from the apex to the baso-lateral angle, at which point it forms a narrow projection: a second ridge, not quite so prominent, runs from the apex to the basal margin, to a point rather nearer to the baso-lateral than to the rostral angle. That part of the valve between this second ridge and the occludent margin has four or five faint longitudinal ridges, whereas the rest of the valve is smooth. Internally there is a deep depression for the adductor muscle, above which the surface is simply concave up to the apex.

what appeared to be upper latera, but as I could not remove them so as to examine their under sides, I am far from sure that they were not broken, angular portions of scuta. If we look to the character of the separate valves, there is a striking and important resemblance between the scuta of *S. cretæ* and *tuberculatum*, in the umbo being seated in a nearly middle point of the occludent margin, and likewise in the two ridges running from the umbo to the baso-lateral angle, and to a central point of the basal margin; in which latter character of the ridges, this species also agrees with *S. semiporcatum*. These facts have determined me, provisionally, to rank the present species under Scalpellum. But on the other hand, if we look to the carina, which, according to our rule, is considered the characteristic valve in this genus, it rather resembles the homologous valve in Pollicipes; for the carina has not any parieties separated from the tectum by a distinct ridge. The terga seldom afford any serviceable generic characters; but as far as they go, they also rather resemble the terga in Pollicipes than in Scalpellum. Hence, it is obvious, that the generic position of *S.* (?) *cretæ* is at present very uncertain.

Valves small, smooth, extremely thin and brittle.

Scutum (fig. 11, *c*); trapezoidal, with the upper part of the valve produced into a sharp point, and with the rostral angle slightly and obliquely cut off. Umbo seated at a little above the middle of the occludent margin, which is straight. The tergal margin is longer than the lateral margin: the basal margin (on the carinal side of the truncated rostral end) forms a right angle both with the lateral and occludent margins. Valve somewhat convex near to the umbo, whence three obscure ridges radiate,—one to the angle between the tergal and lateral margins; a second to the baso-lateral angle, and a third to the bend in the basal margin; these ridges, however, seem to vary in strength, and in the largest specimens could hardly be distinguished: in most of the specimens, the narrow portion of the valve, which ends in the truncated rostral angle, is a little inflected. The lines of growth follow the basal and tergo-lateral margins, and can be traced just bending round the sharp apex, so that a very narrow ledge is added along the upper part of the occludent margin.

Tergum (fig. 21, *a*); sub-rhomboidal, nearly flat: the carinal margin consists of an upper larger portion, and of a lower, shorter portion: the occludent and scutal margins are nearly equal in length. The apex is a little curled towards the scuta, and is sharp; basal angle bluntly pointed. A faint curved ridge runs from the apex to the basal angle, at about one fourth of the entire width of the valve from the carinal margin.

The *Carina* (fig. 11, *b*) widens rapidly downwards from the extremely sharp apex; basal margin spear-shaped, sharply pointed, the two edges meeting each other at about an angle of 75°; exterior surface sub-carinated; in a transverse line the valve is slightly arched, and longitudinally, very slightly bowed inwards: with a lens, traces of longitudinal striæ are visible.

Dimensions. The species seems to have been always small: the largest scutum and tergum were each about a quarter of an inch in length. Probably the individuals were attached in groups to corallines at the bottom of the cretaceous sea.

Genus—POLLICIPES.

POLLICIPES. *Leach.* Journal de Physique, tom. lxxxv, Julius, 1817.[1]
LEPAS. *Linn.* Systema Naturæ, 1767.
ANATIFA. *Brugière.* Encyclop. Méthod. (des Vers), 1792.
MITELLA. *Oken.* Lehrbuch der Naturgesch., 1815.
RAMPHIDIONA. *Schumacher.* Essai d'un Nouveau Syst. &c., 1817 (ante Julium).
POLYLEPAS. *De Blainville.* Dict. des Sc. Nat., 1824.
CAPITULUM (secundum Klein). *J. E. Gray.* Annals of Philos., tom. x, 2d series, Aug. 1825.

Valvæ ab octodium usque ad centum et amplius. Lateribus verticelli inferioris multis; lineis incrementi deorsùm ordinatis. Subrostrum semper adest. Pedunculus squamiferus.

CHARACTERES VALVARUM IN SPECIEBUS FOSSILIBUS.

Carina ab apice ad marginem basalem multum dilatata; apex plerumque liberè prominens; parietes à tecto non distinctè separati; lineæ incrementi parietum parum obliquæ. Scuta plerumque subsolida, convexa, subtrigonalia, margine tergo-laterali plus minusve eminente, sed non angulo in margines duos discreto.

†. *Scuta, aut lævia aut lineis tenuibus incrementi solùm notata.*

A. Scuta, costâ ab apice ad centrum marginis basalis non decurrente.

B. Scuta, costâ, nonnunquam subobsoletâ, ab apice ad centrum marginis basalis decurrente.

††. *Scuta, aut longitudinaliter aut transversè (i. e. secundum lineas incrementi) costata.*

Valves from eighteen to above one hundred in number. Latera of the lower whorl numerous, with their lines of growth directed downwards. Sub-rostrum always present. Peduncle squamiferous.

CHARACTERS OF THE VALVES IN FOSSIL SPECIES.

Carina; widening considerably from the apex, which projects freely, to the basal margin; parietes not distinctly separated from the tectum; lines of growth on the parietes but little oblique. *Scuta* generally somewhat massive, convex, sub-trigonal, with

[1] This is one of the rare cases in which, after much deliberation and with the advice of several distinguished naturalists, I have departed from the rules of the British Association; for it will be seen that *Mitella* of Oken, and *Ramphidiona* of Schumacher, are both prior to *Pollicipes* of Leach; yet as the latter name is universally adopted throughout Europe and North America, and has been extensively used in geological works, it has appeared to me to be as useless as hopeless to attempt any change. It may be observed that the genus *Pollicipes* was originally proposed by Sir John Hill ('History of Animals,' vol. iii, p. 170), in 1752, but as this was before the discovery of the binomial system, by the Rules it is absolutely excluded as of any authority. In my opinion, under all these circumstances, it would be mere pedantry to go back to Oken's 'Lehrbuch der Naturgesch.' for the name *Mitella,*—à work little known, and displaying entire ignorance regarding the Cirripedia.

the tergo-lateral margin more or less protuberant, but not divided by an angle into two distinct margins.

† Scuta smooth, or marked only with fine lines of growth.

A. Scuta without any ridge proceeding from the apex to a nearly middle point of the basal margin.

B. Scuta with a ridge, sometimes faint, proceeding from the apex to a nearly middle point of the basal margin.

††. Scuta either longitudinally or transversely (that is in the direction of the lines of growth) ridged.

As with Scalpellum, the first of the above two paragraphs contains the true generic description, as applicable to recent and fossil species; the second paragraph has been drawn up as an aid in classifying separated valves. This, the most ancient genus of the Lepadidæ, seems also to be the stem of the genealogical tree; for Pollicipes leads, with hardly a break, by some of its species into *Scalpellum villosum;* and Scalpellum leads by Oxynaspis into Lepas and the allied genera: *Pollicipes mitella,* moreover, is nearer allied to the Sessile Cirripedes than is any other Pedunculated cirripede, except, perhaps, Lithotrya, which is also closely connected with Pollicipes. The six recent species of Pollicipes might be divided into three sub-genera: one containing the *P. mitella;* a second, *P. cornucopia, elegans* and *polymerus;* and the third, *P. spinosus* and *serta* (*nov. spec.*) Of the fossil species some, as *P. carinatus, dorsatus, validus,* &c., are related to the first section; others, as *P. reflexus* and *concinnus,* to the second section; and lastly, others, as *P. glaber* and *unguis,* perhaps form a distinct section, though more related to *P. mitella* than to other recent species. As, however, most species are known by only a few of their valves, it is scarcely possible to speak with certainty regarding their finer affinities.

Description: as in the case of Scalpellum, the following remarks are confined to the fossil species of the genus. In all full-grown recent species the number of valves in the capitulum is very large: this seems to have been the case with the Oolitic *P. concinnus,* and probably with most other species, but whether with all may be doubted; from the size of the carinal latera of the lower whorl in *P. unguis,* I suspect that the total number of its valves cannot have been great. The valves are either smooth or plainly marked by the lines of growth, or they rarely have longitudinal ridges, or transverse ridges corresponding to each periodical zone of growth: no recent Pollicipes has a surface of this latter kind. The valves in Pollicipes are often strong and massive, with their apices projecting freely from the capitulum.

Scuta generally three-sided, but sometimes, from either the baso-lateral or rostral angles being truncated, there is an additional lower side. The tergo-lateral margin is either straight or generally more or less convex, but it is never (as far as I know) divided into two distinct margins, as is always the case with Scalpellum owing to the abrupt ending of the upturned lines of growth. The basal margin is either straight or formed of two lines meeting each other at a wide angle, or somewhat irregular. The angle which this basal margin makes with the occludent margin varies much. The occludent margin is slightly arched, and is sometimes exteriorly strengthened by a ledge or rim. A prominent ridge runs in several species from the apex of the valve to the baso-lateral angle; and in another set of species there is a second obscurer ridge running to a nearly middle point of the basal margin: in this latter set, the two ridges no doubt mark the extent to which the rostrum and upper latera overlapped the scutum. Internally there is almost always a deep pit for the adductor scutorum muscle: the upper part of the valve generally projects freely, and is internally marked by lines of growth; sometimes there is a furrow along the upper part either of the occludent or the tergal margin; in the latter case the furrow seems to receive the scuto-occludent angle of the adjoining tergum, and thus locks the two valves together, as in the recent *P. mitella.* In two species the occludent margin at the rostral angle is internally produced downwards into a depending tooth or projection.

Terga: nearly flat, rhomboidal or sub-rhomboidal; a line formed by the converging zones of growth, or a ridge, sometimes steep only on the carinal side, sometimes steep on both sides, runs from the apex to the basal angle. The basal angle is sometimes truncated.

Carina: is either bowed inwards or is straight: it widens from the apex downwards more rapidly than in Scalpellum; generally a considerable upper portion projects freely; this upper portion is always much less concave than the lower part: it is sometimes filled up flat, and sometimes has even a central prominent crest; the basal margin is either bluntly pointed, rounded, or truncated; the parietes are generally more or less inflected, but they are not separated by any defined ridge or angle from the roof or tectum; the lines of growth on the parietes are transverse, or generally only slightly oblique. These characters will, I believe, in nearly all cases serve to distinguish the carina of a Pollicipes from that of a Scalpellum.

Sub-carina: I know of the existence of this valve only in *P. concinnus,* but I cannot doubt that it existed in all, or nearly all, the species. I have sometimes suspected that it might possibly have been absent in *P. unguis* and *glaber,* in which the carinal latera are so large.

Rostrum and *sub-rostrum:* as these valves occur in *P. unguis,* I have little doubt that they are universal; they are apparently present in *P. concinnus;* the *rostrum* always resembles the carina, but is shorter and proportionally broader; a larger proportion, also, seems always to have projected freely, caused no doubt by the more abrupt flexure of this end of the capitulum: this latter character is the most certain one by which the rostrum may be distinguished from the carina. The *sub-rostrum* in *P. unguis* resembles the rostrum, but is smaller, and exteriorly is not carinated.

Upper latera: I know these only in *P. unguis* and *glaber,* in which they consist of a flat triangular plate, and in *P. concinnus,* in which they seem to be diamond-shaped. *Lower latera,* these in *P. concinnus* also seem to be diamond-shaped, as in *P. cornucopia;* in *P. unguis* and *glaber* the apices of these little valves do not project freely, and they have a different appearance from their homologues in any recent species: they are trigonal, with their basal margin rounded and one end produced, to which end a narrow well defined ridge runs obliquely from the apex of the valve.

The peduncle is known only in *P. concinnus;* in this species it is covered with minute quadrangular calcified scales.

†. *Scuta, aut lævia aut lineis tenuibus incrementi solùm notata.*

A. Scuta, costâ ab apice ad centrum marginis basalis non decurrente.

1. Pollicipes Concinnus. Plate III, fig. 1.

Pollicipes concinnus. *J. Morris.* Annals of Nat. Hist., vol. xv, 1845, p. 30, pl. vi, fig. 1, et Mineral Conch., pl. 647, fig. 1.

P. scutis pæne quadratis, margine basali propè rostrum subconcavo, segmento tergo-laterali, è lineis incrementi ut videtur reflexis formato, lato, rotundato et prominente: tergis latis, pæne quadratis: carinæ margine basali, ut videtur acuto.

Scuta, almost square, with the basal margin near the rostrum a little hollowed out; tergo-lateral slip, apparently formed by upturned lines of growth, broad, rounded, and protuberant. *Terga* broad, almost square. *Carina,* with the basal margin apparently pointed.

Oxford Clay, Middle Oolite, attached to an Ammonite. *Mus.* Pearce.

Although to my great regret the state of Mr. Pearce's health has prevented him allowing me to examine the specimens in his possession, yet I have thought it advisable to commence the genus with this species, as it is in a far better state of preservation than any other specimen hitherto discovered. We gain by a single glance the knowledge that at so remote a period as the Middle oolite a true Pollicipes existed. In no other instance that I have heard of, has the peduncle been perfectly preserved. Mr. Morris first named and briefly described this interesting species; subsequently Mr. James Sowerby has given enlarged drawings (without any description) of it in the 'Mineral Conchology;' and it is from these figures that I have drawn up my specific description, which, from this cause, is necessarily imperfect. The figures in this volume are copied from those in the 'Mineral Conchology,' which I may remark have evidently been executed with great care, and Mr. Sowerby's accuracy of observation is universally well known. The peduncle is several times longer than the capitulum: Mr. Morris describes the scales on the peduncle as being small, closely pressed together, somewhat quadrate in form, and each regularly marked by a transverse carinated ridge; this latter character I do not understand. The rostrum is not clearly figured by Mr. Sowerby, but I believe that I can see evidence of its existence. From these materials it would appear that the *P. concinnus* is more nearly related to the recent *P. cornucopia,* and its two nearest congeners, than to the other species of the genus.

2. Pollicipes ooliticus. Tab. III, fig. 2.

Pollicipes ooliticus. *Buckman.* Outline of the Geology of Cheltenham, by Sir R. Murchison, new edit. by James Buckman and H. Strickland, 1845, Tab. iii, fig. 7.

P. scutis triangulis ; superficie undulatâ ; margine basali rectangulè ad marginem rectum tergo-lateralem posito ; segmentum tergo-laterale à lineis reflexis incrementi formatum deest. Carinâ pæne rectâ, semicylindricâ, margine basali quadrato.

Scuta triangular ; surface undulatory ; basal margin at right angles to the straight tergo-lateral margin ; there is no tergo-lateral segment formed by upturned lines of growth. *Carina* nearly straight, semicylindrical, with the basal margin square.

Stonesfield Slate, Lower Oolite : Eyeford. *Mus.* Buckman, and Geolog. Soc.

My materials consist of several scuta, terga, and carinæ, kindly lent me by Professor Buckman, and of another set (which includes the rostrum) presented by him to the Geological Society of London.

Valves : these have a smooth surface, but are undulatory in the direction of the lines of growth ; at the cessation, apparently, of each zone of growth, there was a tendency to form a projecting ridge or plait, as takes place in a far more marked manner in some of the cretaceous species, namely, *P. elegans* and *fallax.* There are also excessively fine, longitudinal striæ, which can be seen only when the valves are held in particular lights ; these seem to have been formed by the so-called epidermis, which we know in the recent *P. mitella* is longitudinally and finely ribbed. *Scuta* (fig. 2, *c*) but slightly convex ; triangular ; basal margin straight, forming a right angle with the tergo-lateral margin, and rather less than a right angle with the slightly arched occludent margin ; the tergo-lateral margin is straight, and not at all protuberant : in the figure the left hand is, as usual, the occludent margin ; I mention this because the valve has a reversed appearance, owing to the unusually small angle which the occludent makes with the basal margin. *Terga* (fig. 2, *d*) rhomboidal, slightly convex, with a rounded ridge, which is central, running from the apex to the broad, rounded basal angle ; the upper carinal and occludent margins stand at right angles to each other, and are short compared to the scutal and lower carinal margins ; there is no trace of a depression parallel to the occludent margin. *Carina* (fig. 2, *a*, *b*) elongated, triangular ; scarcely at all bowed inwards ; not even *sub*-carinated ; basal margin rounded, not at all protuberant. The *Rostrum* differs from the carina only in its greater breadth compared to its length.

Dimensions. The largest scutum is ·6 long, but as there is a broken tergum about 1·1 long, no doubt the species attained a rather large size ; the longest carina is ·7 in length.

Diagnostic characters. This species is best characterised by the straightness of the whole tergo-lateral and of the basal margin of the scuta ; by the ridge being central on the terga ; by the carina not being carinated ; and by the sinuous state of the surface of the valves, intermediate between the smooth species and those with distinct ridges parallel to the zones of growth. The remarkable straightness of the tergo-lateral margin of the scuta is like that in the recent *P. spinosus* and *serta*, and in *Scalpellum villosum*, in all which species, I may observe, the scuta and terga are separated by an interspace of membrane; in these three recent species, however, the basal margin is considerably protuberant. The present species differs apparently from the *P. concinnus* of the Oxford clay, in the basal and tergo-lateral margins of its scuta being straight; in the greater proportional length of the scutal and[1]

[1] 3. POLLICIPES NILSSONII. Tab. III, fig. 11.

POLLICIPES NILSSONII. *Steenstrup.* Kroyer, Naturhist. Tidsskrift, 1839, pl. v, figs. 20—23.

P. scutis triangulis, planis: margine basali cum margine occludente angulum pæne rectum, cum margine recto tergo-laterali, angulum aliquanto minorem formante. Deest segmentum tergo-laterale, lineis incrementi reflexis formatum. Carinâ introrsùs admodùm arcuatâ, crassâ; marginis basalis mucrone obtuso.

Scuta triangular, flat; basal margin forming nearly a rectangle with the occludent margin, and a somewhat lesser angle with the straight tergo-lateral margin. There is no tergo-lateral slip formed by upturned lines of growth. Carina much bowed inwards, massive, with the basal margin bluntly pointed.

Scania (Balsberg, Kopinge, Ffo., &c.) *Mus.* Univers. Copenhagen.

Professor Steenstrup has described, under the name of *Pollicipes Nilssonii*, a large carina, and apparently a sub-carina and rostrum, and he remarks that these perhaps belong to the same species with the terga, named by him *P. undulatus.* M. Angelin, however, believes that the latter belong to the species already described as *Scalpellum solidulum.* With the specimens of the present species, M. Angelin has lately found three small scuta, which he believes belonged to it. These scuta are so extremely worn, that I should not have ventured to have named them, had it not been advisable to give figures of the remarkable carina already named as *P. Nilssonii.* Should it hereafter be proved that the following scuta belong to some other carina, then a new name will have to be attached to them.

Scuta (fig. 11, *a*) flat, thick, triangular, not much acuminated; basal margin forming almost a rectangle with the occludent margin; tergo-lateral margin (in present condition) straight, forming a rather less angle with the basal than does the occludent margin. There is no trace of a slip or portion of valve along the tergo-lateral side, formed by upturned lines of growth. Internally, the pit for the adductor muscle is deep; the central portion of the apex above the pit is prominent; apparently there was no internal furrow. Length of longest specimen only ·4 of an inch.

Carina (fig. 11, *b*, *c*) strong, with the upper part unusually massive; though in a worn condition, there are distinct traces of its having been longitudinally and slightly ribbed. Strongly carinated, the two arched sides meeting each other at about a rectangle; much bowed inwards, and widening much from the apex to the base; upper portion, about one fifth of the entire length of the carina, seems (for the worn condition prevents certainty) to have projected freely; beneath the upper freely projecting portion, the inner margins are nearly straight; the depth of the shell, measured from the central crest to the inner margin, is, in the lower half, remarkably great, and consequently the valve in the same part is internally concave to a remarkable depth; the upper freely projecting portion is only slightly concave, and is singularly massive, from having been filled up with solid shelly layers. The basal margin is bluntly pointed, the edges meeting each other at about a right angle; in the lower part of the valve the lines of growth are of course parallel to the basal edges, but higher up they meet at a more open angle, and consequently the carina of a young individual must have had its basal margin less projecting. When the sides of the carina are examined carefully, a portion, about one fourth of its entire depth, can be observed to lie a very little more inwardly inflected than the more central part, so as not to form quite a continuous surface with lower carinal margins compared with the upper carinal and occludent margins of the terga, and lastly in the basal margin of the carina being truncated; it differs from *P. planulatus* of the Oxford Clay, and therefore its other nearest relative in age, by the basal angle of the terga being rounded, instead of square as in that species.

the two broad arched roof-sides; and in these slips the lines of growth run almost parallel to the inner margin of the valve: in this respect the valve approaches in character to that of *Scalpellum*. The heels or baso-lateral angles apparently projected slightly, as I infer from a slight downward curvature in the lines of growth, along a line corresponding with the heel, and separating the roof-part from the inflected walls of the carina.

Sub-carina (fig. 11, *d*): in Professor Steenstrup's collection there are several worn valves which appear to have been sub-carinæ; in shape approximately semi-conical; the basal margin being almost semi-oval, with the two corners a little inflected; hence the valve is deeply concave to an unprecedented degree, and this is quite conformable with the singular sectional outline of the carina (*c*). About one fourth part of the length of the valve must have projected freely; the outer surface is longitudinally ribbed, and the lines of growth remarkably undulatory.

Rostrum (fig. 11, *e*): this valve which I believe to be the rostrum resembles the sub-carina, but is more open, less high, and with a larger proportion, namely half, of its entire height freely projecting; the semi-oval basal margin is slightly sinuous, the projecting points corresponding with the external longitudinal ribs.

Length of carina, 1·5; of the largest of the sub-carinæ, ·6; of the largest rostrum, ·45 of an inch.

4. Pollicipes Hausmanni. Tab. III, fig. 3.

Pollicipes Hausmanni. *C. L. Koch* and *Dunker*. Norddeutsch. Oolithgebilder, p. 52, Tab. vi, fig. 6.
— — *F. A. Roemer*. Versteinerung. Norddeutsch. Oolithengebirges, p. 211, Tab. iv, fig. 2.

P. scutis subtriangulis, angulo baso-laterali valde rotundato; apice producto; margine basali cum margine occludente angulum pæne rectum formante; internâ apicis superficie prominente, margineque tergali sulcato.

Scuta, subtriangular, with the baso-lateral corner much rounded, and with the apex produced; basal margin forming nearly a right angle with the occludent margin; apex with its internal surface prominent, and with the tergal edge furrowed.

Hilsthon, des Elligser Brinkes. (Lower Greensand, Germany.)

Messrs. Koch and Dunker have given a full and detailed account of this species, together with truly excellent figures, and I have nothing to add to their remarks, but will re-describe, for the sake of uniformity, the valves of this species, which, through the kindness of Professor Steenstrup and Professor Dunker, I have examined. The valves are slightly worn. The figures given in tab. III are not, I think, so good as most of the others.

Scuta (Tab. III, fig. 3, *b*, *c*) moderately convex, sub-triangular; apex much acuminated, slightly curved towards the terga; surface smooth, faintly marked with zones of growth, and, especially near the apex, with faint lines and furrows radiating from it. There is no distinct ridge proceeding from the apex to the baso-lateral angle, which is so much rounded that the basal margin blends into the tergo-lateral; it must, however, be remarked, that the specimens are worn. The occludent margin stands at right angles to the basal; and the lower part of the tergo-lateral margin forms rather above a right angle with it. Internally (*c*), there is a deep pit for the adductor scutorum, and in the upper part, close to the tergal margin, a deep furrow; the central portion is prominent; the occludent margin keeps nearly of the same thickness up to the apex of the valve.

Terga (Tab. III, fig. 3, *d*), nearly flat, sub-rhomboidal, or rather pointed oval, with the scutal half

5. Pollicipes politus. Tab. III, fig. 4.

P. scutis ferè rhombicis, lævissimis; margine basali cum margine occludente angulum recto majorem formante; margine occludente projecturâ parietali,[1] *lineari, minutâ instructo; internâ apicis superficie concavâ.*

Scuta, almost rhomboidal, excessively smooth, basal margin forming above a right angle with the occludent margin, which latter is exteriorly furnished with a linear, minute, wall-sided ledge; apex with its internal surface concave.

Mus. Bowerbank. Locality and formation unknown; from the state of another specimen fastened on the same board, I think probably from the Gault; the colour of the substance in the cracks of the valve countenances this same opinion.

I have been unwilling to fix a specific name to a single, much broken scutum; but as even in its present state it can be clearly seen to be distinct, and as this is the typical valve in this genus, I have felt myself compelled to do so.

Scutum sub-rhomboidal, approaching to oval in outline: rather thin; surface excessively smooth; slightly convex, but with a narrow portion along the occludent margin, somewhat inflected: exteriorly close to this same margin, or rather almost forming it, (*b*) there is an extremely narrow, sharp, wall-sided, projecting ledge. The occludent margin is slightly arched, and forms, with the basal margin, an angle considerably above a right angle, so that the whole baso-lateral corner of the valve is much produced: the lower part of the tergo-lateral margin is at right angles to the basal margin. Baso-lateral angle smoothly rounded, with no trace of a ridge running from it to the apex, though this is the line of chief flexure of the valve. Internally, the valve has been much injured; the de-

protuberant; surface smooth, but near the pointed, slightly curled apex, it is marked by fine radiating lines; carinal margin regularly curved from the apex to the basal angle, which latter is not very sharp. A curved ridge (formed by the surface of the shell being lower on the carinal than on the other side) connects the upper and basal apices, running almost parallel to the carinal margin, and at about one-fourth of the entire width of the valve from the latter margin. Occludent margin shorter than the scutal; rounded, protuberant, with a depression parallel to it; the scutal margin, corresponding with this depression, being slightly hollowed out; a small portion of the apex projects freely. Internally, and nearer to the occludent than to the carinal margin, there are three or four short parallel longitudinal ridges or crests, as described in *Scalpellum arcuatum.*

Carina (Tab. III, fig. 3, *a*) moderately bowed inwards, widening gradually from the apex to the basal margin, which is rounded and protuberant, and with a trace of an angular bend in the middle; exteriorly the surface presents just a trace of being sub-carinated; roof convex; the upper part of the valve projects freely.

Rostrum: Koch and Dunker figure valves, which, from their general appearance, breadth, and apparently large proportional upper, freely-projecting portion, I have little doubt have been rightly considered by them as rostra; they are, however, longitudinally plicated or striated to a greater extent than the other valves.

[1] Parietali, *i. e.* lateribus utrinque perpendicularibus.

each scale lies exactly between two scales in the whorls, both above and below: this is, in fact, the case with the large lateral scales in Loricula, but the ends of the scales in the same whorl, instead of, as is usual, quite or nearly touching each other, are here far removed from each other, so that each whorl is broken by wide open spaces. In the marked difference in size between the lateral scales and those in the two end rows;—in the latter scales not intersecting each other, but presenting a straight, medial, rostral and carinal suture;—and lastly, in each alternate whorl having a different number of scales, namely, four in one and six in the other, Loricula differs from every other known Cirripede.

INDEX.

[N.B.—The names in italics are either synonyms or doubtful species.]

INDEX.

TAB. I.

[*In every case right-hand Scuta and Terga are figured; hence the occludent margins are always to the left-hand.*]

Fig. 1. Scalpellum magnum:—fig. (*a*) natural size, the rest magnified twice.

(*a*) Imaginary restored figure, of natural size.
(*b*) Carina, dorsal view.
(*c*) Scutum.
(*d*) Tergum.
(*e*) Upper latus.
(*f*) Carina, lateral view.
(*g*) Rostral latus, narrow variety.
(*h*) Rostral latus, inside view of, broad variety.
(*i*) Ditto, outside view of, narrow variety.
(*k*) Ditto, seen in profile.
(*l*) Carinal latus, outside view.
(*m*) Ditto, inside view.
(*n*) Ditto, outside view, broad variety.

Fig. 2. Scalpellum angustum:—carina, dorsal and lateral views of, copied from Dixon's 'Geology and Fossils of Sussex,' Tab. xxviii, fig. 9.

Fig. 3. Scalpellum quadratum, fig. (*a*) natural size, the rest magnified twice.

(*a*) Specimen as found embedded, with the valves in nearly their natural positions; the end of the rostral latus ought to touch the short basal side of the upper latus.
(*b*) Scutum.
(*c*) Tergum.
(*d*) Carina, lateral view.
(*e*) Upper latus.
(*f*) Rostral latus.
(*g*) Carinal latus.
(*h*) Scutum, internal view.
(*i*) Carina, dorsal view.
(*k*) Ditto, section across middle of valve.

Fig. 4. Scalpellum fossula; all the figures magnified twice, except (*h*), which is four times the natural size.

(*a*) Scutum.
(*b*) Tergum.
(*c*) Carina, side view.
(*d*) Upper latus.
(*e*) Carinal latus.
(*f*) Scutum, internal view.
(*g*) Carina, dorsal view.
(*h*) Ditto, section across middle.

Fig. 5. Scalpellum trilineatum; all the figures magnified twice, except (*c* and *d*), which are four times the natural size.

(*a*) Carina, dorsal view.
(*b*) Ditto, lateral view.
(*c*) Ditto, section of, lower part.
(*d*) Carina, section of, near apex.
(*e*) Tergum.

Fig. 6. Scalpellum semiporcatum; scutum, magnified three times.

Fig. 7. Scalpellum arcuatum; all the figures magnified twice, except (*g*), which is four times the natural size.

(*a*) Carina, dorsal view.
(*b*) Ditto, lateral view.
(*c*) Tergum, inside view.
(*d*) Ditto, outside view.
(*e*) Surface of carina, much magnified.
(*f*) Scutum.
(*g*) Section across carina.

Fig. 8. Scalpellum solidulum; natural size, except (*f*).

(*a*) Tergum.
(*b*) Carina, dorsal view.
(*c*) Ditto, internal, almost lateral, view.
(*d*) Carina, section of upper part.
(*e*) Carinal latus, natural size.
(*f*) Ditto, much magnified.

Fig. 9. Scalpellum simplex; twice natural size.

(*a*) Carina, dorsal view. (*b*) Carina, lateral view. (*c*) Carina, section of.

Fig. 10. Scalpellum tuberculatum; largely magnified.

(*a*) Tergum.
(*b*) Carina, dorsal view.
(*c*) Ditto, lateral view.
(*d*) Scutum; (*e*) basal margin, (*f*) occludent margin; these two margins ought to form a rather larger angle.

Fig. 11. Scalpellum (?) cretæ; largely magnified.

(*a*) Tergum. (*b*) Carina. (*c*) Scutum.

Tab I

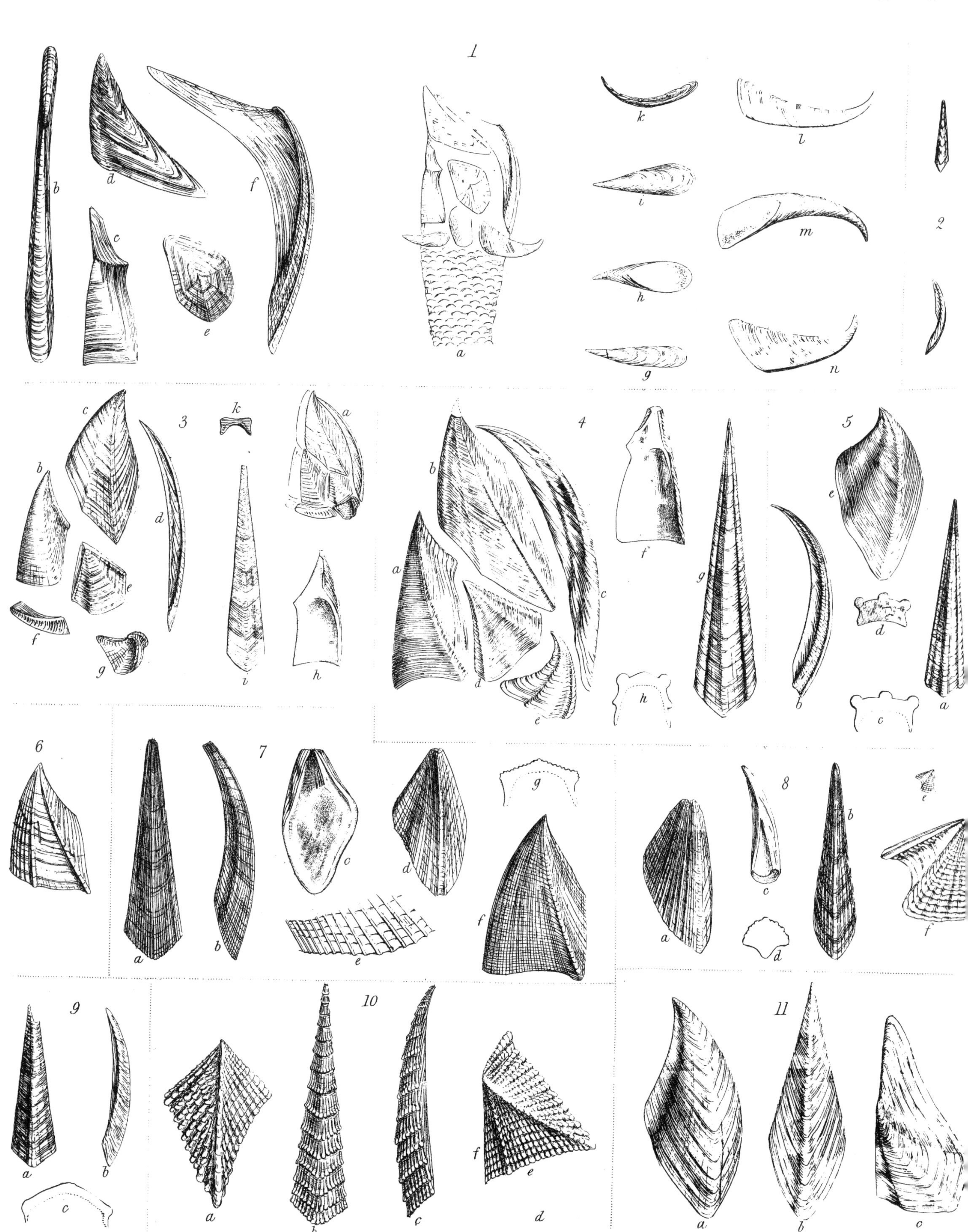

TAB. II.

[*In every case right-hand Scuta and Terga are figured; hence the occludent margins are always to the left-hand.*]

Fig. 1. Scalpellum maximum, *var.* typicum.

(*a*) Carina, twice natural size.
(*b*) Ditto, lateral view, natural size.
(*c*) Ditto, ditto, twice natural size.
(*d*) Section of carina across middle of valve, twice natural size.
(*e*) Section of carina across lower part of valve.

Fig. 2. Scalpellum maximum, *var.* cylindraceum; all the figures twice the natural size, except the sections.

(*a*) Carina.
(*b*) Ditto, lateral view.
(*c*) Section of carina, upper part.
(*d*) Ditto, ditto, lower part.

Fig. 3. Scalpellum maximum, *var.* sulcatum.

(*a*) Carina, twice natural size.
(*b*) Ditto, natural size.
(*c*) Ditto, lateral view.
(*d*) Section across the middle of carina, four times natural size.

Fig. 4. Scalpellum maximum; carinal latus, two varieties.

(*a*) Natural size.
(*b*) Magnified four times.
(*c*) Natural size.
(*d*) Magnified four times.

Fig. 5. Scalpellum maximum; Tergum, *Var.* I., natural size and twice magnified.

Fig. 6. Scalpellum maximum; Tergum, *Var.* II., natural size and twice magnified.

Fig. 7. Scalpellum maximum; Tergum, *Var.* III.

(*a* and *a*) Natural size, and twice magnified.
(*b*) Internal view, twice magnified.

Fig. 8. Scalpellum maximum; *Var.* I.

(*a*) Scutum, twice natural size: some longitudinal lines have been erroneously introduced in this engraving.
(*b*) Scutum, natural size.
(*c*) Ditto, internal view, twice natural size.

Fig. 9. Scalpellum maximum; *Var.* II.

(*a*) Scutum, outside view, twice natural size.
(*b*) Scutum, internal view, twice natural size.
(*c*) Scutum, outside view, natural size.

Fig. 10. Scalpellum maximum; *Var.* III. Scutum, inside view, natural size.

Fig. 11. Scalpellum lineatum.

(*a*) Scutum, natural size.
(*b*) Scutum, twice natural size.

Fig. 12. Scalpellum lineatum; Carina twice natural size.

Fig. 13. Scalpellum hastatum.

(*a*) Carina, twice natural size.
(*b*) Ditto, natural size, lateral view.
(*c*) Carina, lateral view, twice natural size.
(*d*) Ditto, section across middle of valve.

Tab. II.

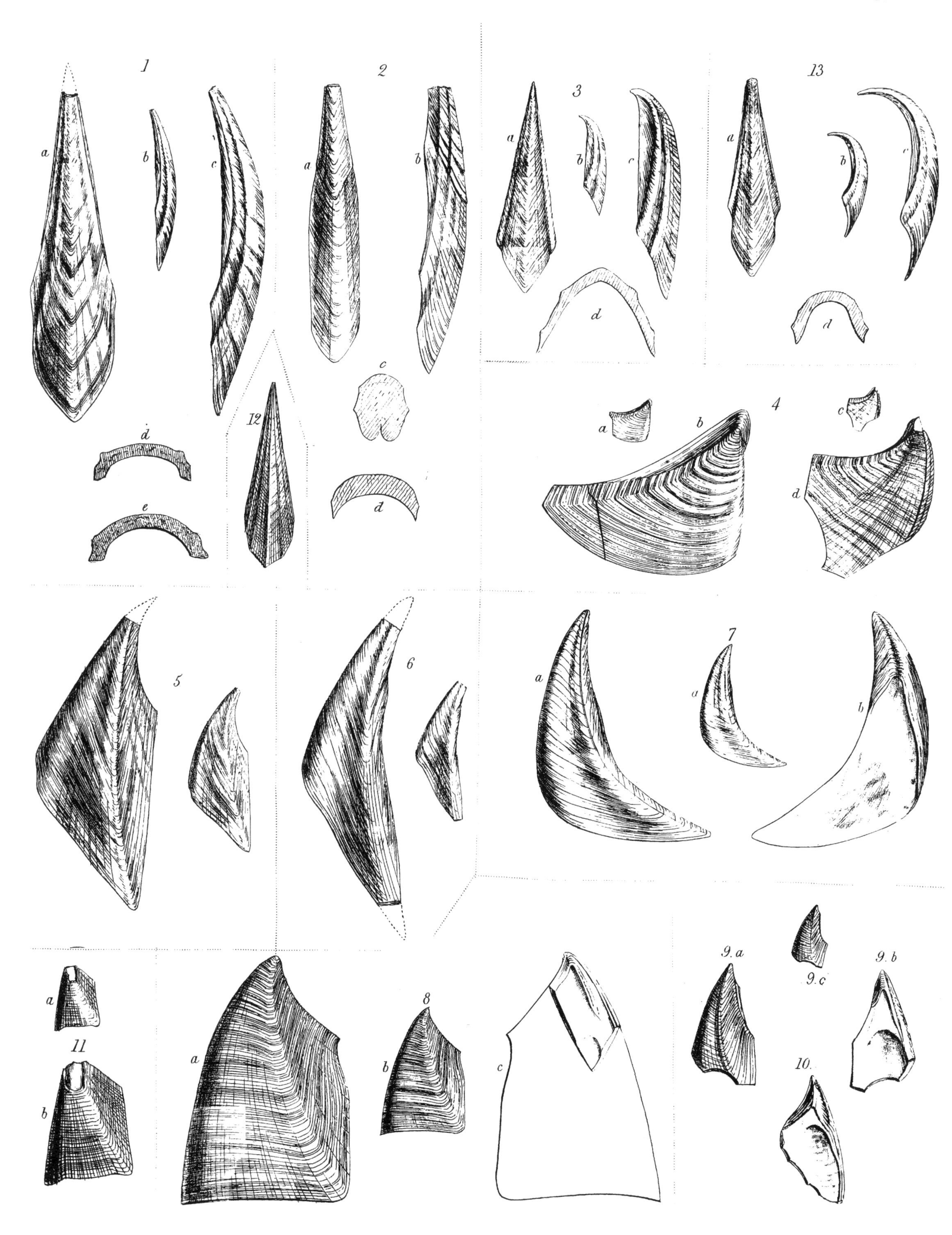

1
a
b
c
d
e
2
a
b
c
d
12
3
a
b
c
d
13
a
b
c
d
4
a
b
c
d
5
6
7
a
a
b
11
a
b
8
a
b
c
9. a
9. b
9. c
10.

TAB. III.

[*In every case right-hand Scuta and Terga are figured; hence the occlu ent margins are always to the left-hand.*]

Fig. 1. Pollicipes concinnus; copied from the Mineral Conchology, Pl. 647.

(*a*) Group of specimens as found adhering to an Ammonite, of the natural size.
(*b*) Capitulum enlarged.
(*c*) Scales of the peduncle magnified.

Fig. 2. Pollicipes ooliticus.

(*a*) Carina.
(*b*) Rostrum.
(*c*) Tergum.
(*d*) Scutum.

Fig. 3. Pollicipes Hausmanni.

(*a*) Carina.
(*b*) Scutum.
(*c*) Scutum, inside view of.
(*d*) Tergum.

Fig. 4. Pollicipes politus.

(*a*) Scutum.
(*b*) Small portion of the occludent margin, much magnified, to show the narrow prominent ledge.

Fig. 5. Pollicipes elongatus.

(*a*) Tergum, about half natural size.
(*b*) Scutum, much magnified; the impression on the chalk gives the general outline.
(*c*) Scutum, natural size.

Fig. 6. Pollicipes acuminatus.

(*a*) Scutum, outside view, figure restored.
(*b*) Inside view of actual specimen.

Fig. 7. Pollicipes Angelini.

(*a*) Scutum.
(*b*) Ditto, inside view of.
(*c*) Tergum.
(*d*) Section across the middle of the tergum, to show form of surface.

Fig. 8. Pollicipes reflexus.

(*a*) Carina, dorsal view.
(*b*) Ditto, lateral view.
(*c*) Ditto, section beneath the middle.
(*d*) Tergum.
(*e*) Scutum; the letter (*e*) stands close to the occludent margin.
(*f*) Upper latus.

Fig. 9. Pollicipes carinatus.

(*a*) Tergum.
(*b*) Ditto, internal view.
(*c*) Carina.
(*d*) Scutum.
(*e*) Scutum, inside view of.
(*f*) Carina, section of, near apex.
(*g*) Rostrum, inside view.
(*h*) Ditto, lateral view.
(*i*) Ditto, dorsal view.

Fig. 10. Pollicipes glaber.

(*a*) Tergum, broken carina, upper latus, and a latus of the lower whorl; all these valves belonged to the same individual, and show their relative natural sizes.
(*b*) Scutum, natural size, of another individual.
(*c*) Ditto, ditto, four times magnified.
(*d*) Ditto, inside view of upper part; the left-hand is the occludent margin.
(*e*) Tergum, four times magnified.
(*f*) Carina.
(*g*) Ditto, small portion of lateral margin, close above the basal margin, much enlarged.
(*h*) Rostrum, natural size.
(*i*) Ditto, four times magnified.
(*k*) Upper latus, three times magnified.
(*l*) Latus (probably from near the rostrum) of the lower whorl, three times magnified.

Fig. 11. Pollicipes Nilssonii.

(*a*) Scutum, from a small worn specimen.
(*b*) Carina, natural size, lateral view of.
(*c*) Ditto, section of, across the middle.
(*d*) Sub-carina, inside view, natural size.
(*e*) Rostrum, inside view, natural size.

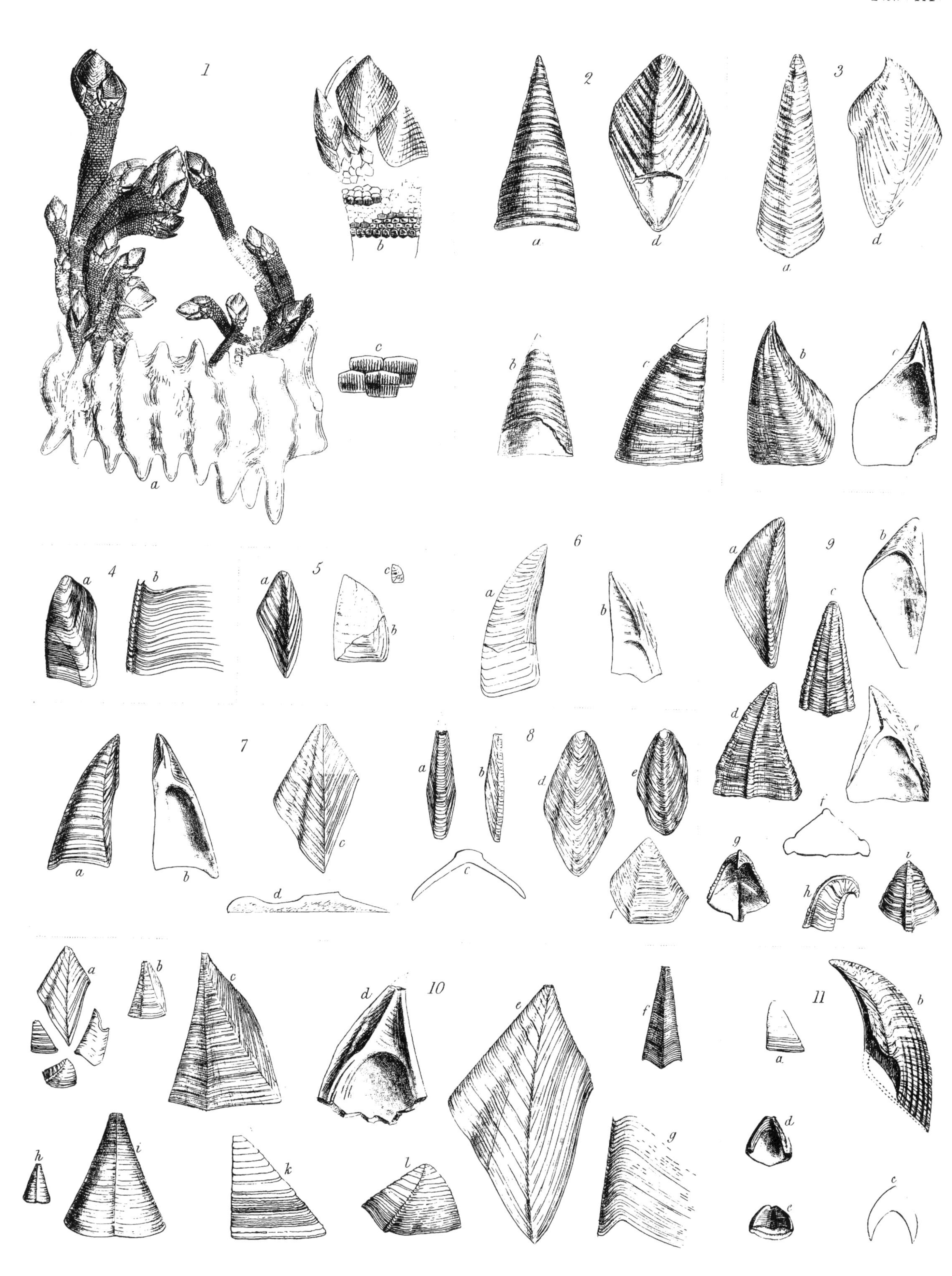
1
a
b
c
2
a
d
b
c
3
a
d
b
c
4
a
b
5
a
b
c
6
a
b
7
a
b
c
d
8
a
b
c
d
e
f
9
a
b
c
d
e
f
g
h
i
10
a
b
c
d
e
f
g
h
i
k
l
11
a
b
c
d
e

TAB. IV.

[*N.B.—A left-hand scutum (fig. 4, f), and a left-hand tergum (fig. 5, b), have been accidentally figured instead of right-hand valves, as in all the other Figures and Plates; hence these two hold reversed positions, and in both the occludent margins are to the right, instead of to the left hand.*]

Fig. 1. Pollicipes unguis; all the figures except (*d*) are from the same individual, and are magnified twice: (*d*) is of the natural size.

(*a*) Carina.
(*b*) Tergum, inside view.
(*c*) Ditto, outside view.
(*d*) Ditto, ditto, (var., natural size.)
(*e*) Rostrum.
(*f*) Sub-rostrum.
(*g*) Upper latus.
(*h*) Latus of the lower whorl, probably adjoining the carina.
(*i*) Ditto, inside view.
(*k*) Latera of the lower whorl, probably the two adjoining the rostrum.
(*l*) Ditto, inside view of.

Fig. 2. Pollicipes validus; all figures natural size.

(*a*) Carina.
(*b*) Ditto, inside view.
(*c*) Ditto, lateral internal view of.
(*d*) Ditto, section of upper part.
(*e*) Scutum.
(*f*) Ditto, inside view of.
(*g*) Ditto, inside view of another specimen.

Fig. 3. Pollicipes gracilis.

(*a*) Scutum, natural size.
(*b*) Scutum, magnified, inside view.

Fig. 4. Pollicipes dorsatus; all figures natural size, except (*d*), magnified twice.

(*a*) Carina, inside view of.
(*b*) Ditto, outside view of.
(*c*) Ditto, section of upper part of.
(*d*) Tergum, magnified twice.
(*e*) Tergum, natural size.
(*f*) Scutum, left-hand valve (*vide supra*).

Fig. 5. Pollicipes striatus; figures natural size, and magnified twice.

(*a*) Carina.
(*b*) Tergum; a left-hand valve (*vide supra*).
(*c*) Scutum.

Fig. 6. Pollicipes semilatus; magnified about ten times.

Fig. 7. Pollicipes rigidus; all figures thrice natural size, except (*f*), which is twice

(*a*) Carina.
(*b*) Ditto, side view of.
(*c*) Ditto, section of lower part of.
(*d*) Scutum.
(*e*) Tergum.
(*f*) Scutum, inside view.

Fig. 8. Pollicipes fallax; figures twice natural size.

(*a*) Scutum.
(*b*) Tergum.

Fig. 9. Pollicipes elegans.

(*a*) Carina, thrice natural size.
(*b*) Ditto, section of.
(*c*) Scutum, twice natural size.
(*d*) Tergum, twice natural size.

Fig. 10. Pollicipes Bronnii; figures magnified twice.

(*a*) Carina.
(*b*) Ditto, lateral view of.
(*c*) Ditto, section of, near basis.
(*d*) Carina, section of, at one third of length fr the apex.

Fig. 11. Pollicipes planulatus; Tergum, upper figure, natural size.

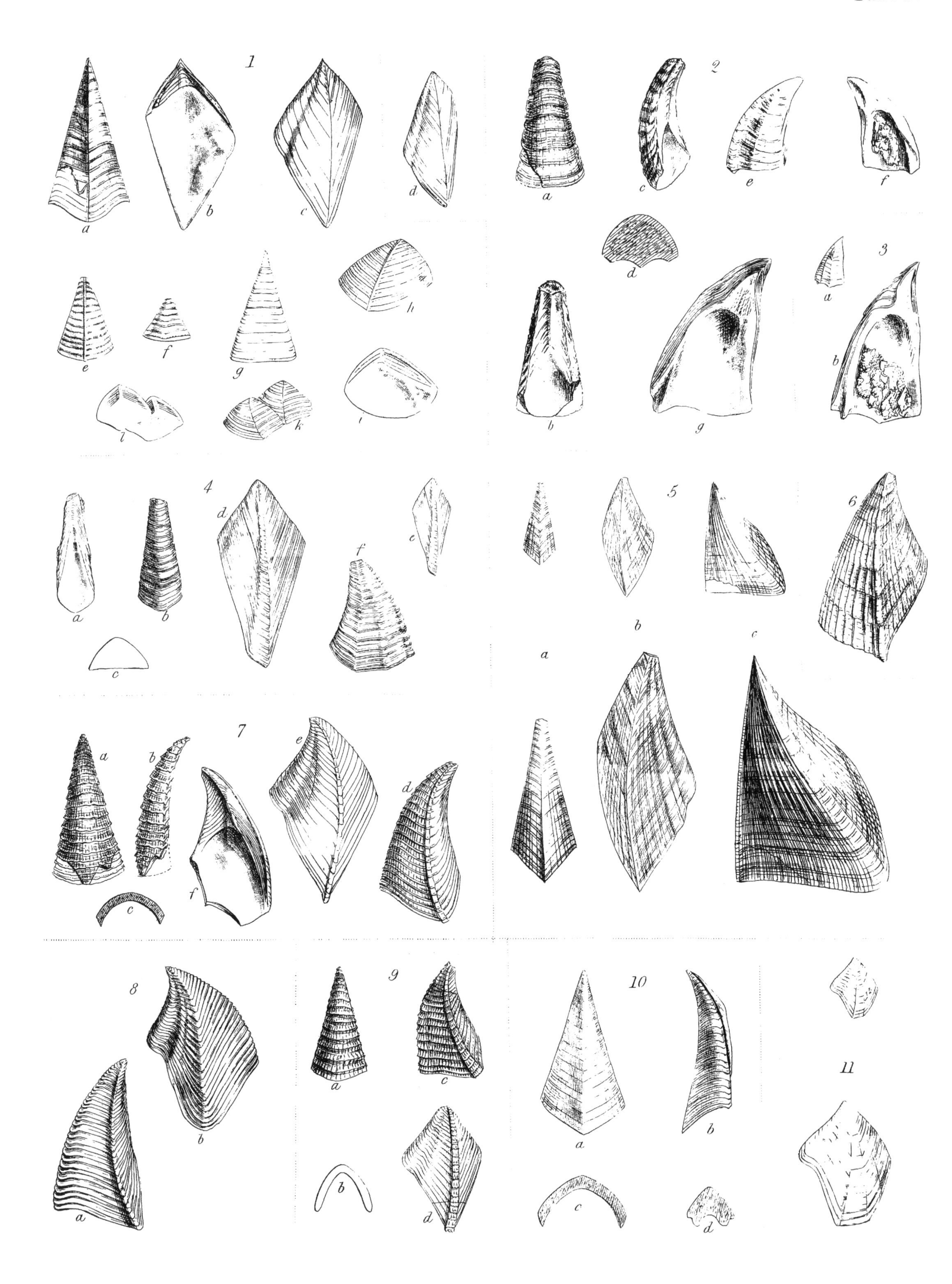
1
a
b
c
d
e
f
g
h
i
k
l
2
a
b
c
d
e
f
g
3
a
b
4
a
b
c
d
e
f
5
a
b
c
6
7
a
b
c
d
e
f
8
a
b
9
a
b
c
d
10
a
b
c
d
11

TAB. V.

Fig.

1. Loricula pulchella ; natural size, as found embedded.

2. Ditto — Left-hand main valve, and scales of the peduncle, magnified three times.

3. Ditto — Right-hand main valve (scutum), and scales of the peduncle, magnified three times.

4. Ditto — Imaginary, restored figure.

Tab. V

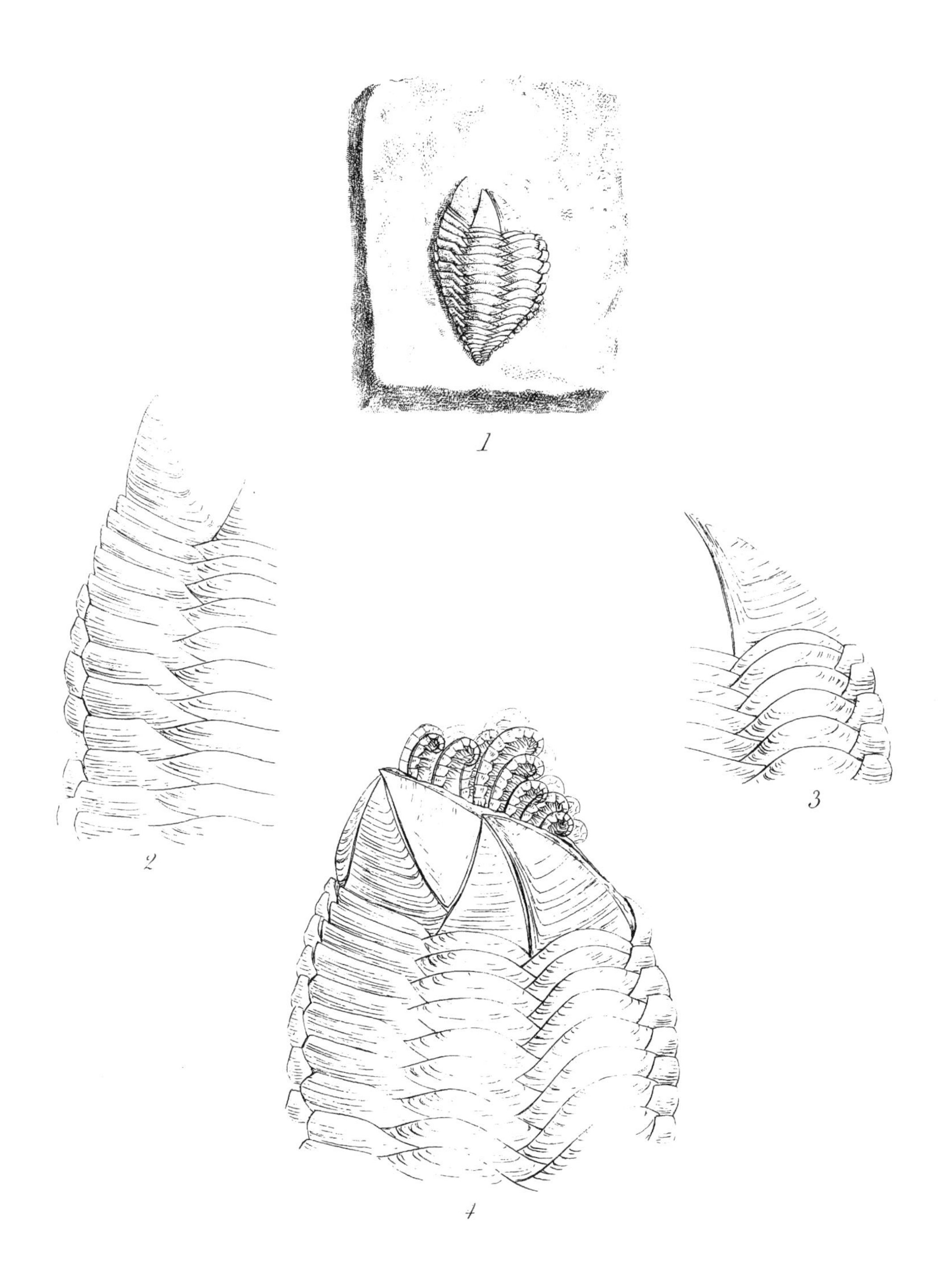

THE

PALÆONTOGRAPHICAL SOCIETY

INSTITUTED MDCCCXLVII.

LONDON:

MDCCCLIV.

A MONOGRAPH

ON THE

FOSSIL BALANIDÆ

AND

VERRUCIDÆ

OF

GREAT BRITAIN.

BY

CHARLES DARWIN, F.R.S., F.G.S.

LONDON:

PRINTED FOR THE PALÆONTOGRAPHICAL SOCIETY.

1854.

PREFACE.

As the present short Monograph completes my work on British Fossil Cirripedes, I beg to be permitted again to have the satisfaction of returning my very sincere thanks to the many naturalists who have placed their collections at my disposal, and have given me the freest permission to use the specimens, in whatever manner I might find necessary. —My thanks are most especially due to Mr. Searles Wood, Mr. Bowerbank, and Sir Charles Lyell; and to Mr. J. de C. Sowerby for the use of the original specimens figured in the 'Mineral Conchology.' I lie, also, under much obligation to Mr. D. Sharpe, Mr. Greenough, Mr. Smith of Jordan Hill, Professor Tennant, Mr. Charlsworth, Mr. F. Edwards, Dr. T. Wright, Professor Forbes, Professor Henslow, M. Bosquet, and to many others. I must, also, be permitted to tender my grateful thanks to the Council of the Palæontographical Society for the very liberal manner in which they have allowed my two Monographs to be illustrated.

INTRODUCTION.

Cirripedia may be divided, as I have recently shown in a monograph on the Balanidæ published by the Ray Society, into three Orders: of these, the Thoracica includes all ordinary Cirripedes, and all ever likely to be found fossil, and therefore the two other orders may be here passed over without notice. The Thoracica contains three Families: the Balanidæ or Sessile Cirripedes, which in a recent state so abound on the shores of almost every quarter of the world, and which are so frequently found in Tertiary deposits; the Verrucidæ, which includes only a single genus very singular from its asymmetrical shell; and the Lepadidæ, or Pedunculated Cirripedes; of the latter, the Fossil species have been already published by the Palæontographical Society. The Balanidæ and Verrucidæ will be treated of in the following pages. As yet only sixteen species in these two families have been found fossil in Great Britain; and of these sixteen, nine are still living forms. As the latter are known only imperfectly in their fossil condition, and as they have lately been described by me in full detail, I have thought it best here only to make a few remarks on such portions of the shells of each species as have hitherto been discovered, adding a few illustrations, such as appeared to me desirable. The extinct species will be fully described: of these, all the figures given are from British specimens. But of the species found both living and fossil, I have in several instances (always so stated) given drawings from recent specimens; some of the valves either not having been found fossil, or found only in an imperfect and not characteristic condition. As so few species in the several genera are known in a fossil condition, I have thought it quite superfluous to give long generic descriptions, which would have required constant references to many species exclusively found living.

In my former monograph on the Fossil Lepadidæ, I remarked how much the natural history of Cirripedes has been neglected; and this remark is eminently applicable to the Balanidæ, or Sessile Cirripedes. Even the British recent species have not been well made out, and as for the fossil species, scarcely anything has been done, besides the publication of some figures, in very few instances accompanied by the details which are absolutely necessary for the identification of the species.

Owing to the great variation in external characters, to which almost all the species are subject, and likewise in the case of the principal genus, Balanus, to its being a very natural genus, that is, to the species following each other in close order, it is not easy to exaggerate the difficulty of identifying the species, except by a deliberate examination of the internal and external structure of each individual specimen. Every one who has collected Sessile Cirripedes must have perceived to what an extent their shape depends on their position and grouping. The surface of attachment has a great effect on that of the shell; for as the walls are added to at their bases, every portion has at one time been in close contact with the supporting surface; thus I have seen a strongly-ribbed species (*B. porcatus*) and a nearly smooth species (*B. crenatus*) closely resembling each other and both having a peculiar appearance, owing to their having been attached to a pecten. Dr. Gray has pointed out to me specimens of the recent *B. patellaris*, curiously pitted like the wood to which they had adhered; and numberless other instances might be added. Quite independently of the effect produced by the surface of attachment, the degree to which the longitudinal folds and ribs are developed on the parietes, is variable in most of the species, as in *B. tintinnabulum* and even in *B. porcatus;* the presence or entire absence of these ribs often surprisingly alters the whole aspect of the shell. The persistence of the so-called epidermis is in some degree variable, though this is of little importance in regard to fossil specimens. Again, some species in certain localities are all subject to the disintegration of the entire outer lamina of the walls; and in such cases (as with *B. perforatus*) there is not the smallest resemblance between the corroded and perfect specimens. The size of the orifice, and consequently of the operculum, compared with the shell, varies accordingly as the shell is more or less conical or cylindrical; in the latter case, the summits of the radii are generally more oblique and the orifice consequently more deeply toothed than in the more conical varieties. Size is a serviceable character in some cases, but very many specimens are required to ascertain the average or maximum size of each species, for there is no method of distinguishing a half-grown from a full-grown specimen; and I believe, as long as the individual lives, so long does it go on growing. *Colour* is of very considerable service; but the majority of the species have their white or nearly white varieties, the latter being sometimes as numerous as the coloured ones.

Besides the slight variation in the obliquity of the summits of the radii and alæ, dependent on the more or less cylindrical form of the shell, in some species, as in *B. tintinnabulum* and *porcatus*, their obliquity also varies occasionally from unknown causes, and thus greatly affects the general appearance of the shell. In some few species, as in *B. perforatus*, the radii are often either not at all developed, or are of very variable width; in others, when the shell has become cylindrical, or when very old, the radii cease to grow, and from the disintegration of the whole upper part of the shell, with the continued growth of the lower part, the radii at last come to exist as mere fissures: I have seen instances of this in *B. tintinnabulum* and *porcatus*. Nevertheless, the obliquity of the upper margin, and the breadth of the radii are useful characters; and still more useful is the fact whether the upper margins are smooth and arched, or straight and jagged. The fact of the terga being more or less beaked is useful: as is, likewise, the presence of striæ, or furrows, or rows of pits, radiating from the apices of the scuta; but to ascertain the presence of these marks, it is almost invariably necessary to clean and examine the scuta with a lens; these ridges and furrows, moreover, in some species, as is strikingly the case with *B. tintinnabulum*, and in less degree with *B. concavus*, appear and disappear, and vary without any apparent cause.

Now if we reflect that form, size, state and nature of the surface, presence of epidermis, relative size of the orifice, presence of longitudinal ribs, tint, and often the existence of any colour, are all highly variable in most of the species; and that the obliquity of the summits of the radii, and the presence of longitudinal striæ on the scuta, are variable in some of the fossil species, we shall perceive how difficult it must ever be to distinguish the species from external characters. Let no one attempt to identify the species of this genus, without being prepared to separate, clean, and carefully examine with a microscope the basis and parietes, and both the under and upper surfaces of the opercular valves; for I feel convinced, that he will otherwise throw away much labour. Moreover, in many cases, it is almost necessary, on account of the variability of the characters, to possess several specimens. From these facts, I have not hesitated to use characters, for the identification of the species, which require close examination, though I would gladly have seized on external characters could I have found such even moderately constant.

The least varying, and therefore most important characters, must be taken from the internal structure of the parietes, radii, and basis: not that these characters are absolutely invariable; thus the porosity of the parietes is slightly variable in the recent *B. glandula*, and highly variable in the fossil *B. unguiformis*. The porosity of the basis is in some degree variable in *B. spongicola*. Characters derived from the general shape, and from the ridges and depressions on the under side of the scuta and terga, especially of the scuta, are highly serviceable; though even these are variable. The cause of the opercular valves offering more useful characters, as far as outline is concerned, than do the walls of the shell, is no doubt due to their being almost independent of any influence from the nature of the surface of attachment. Even the ridges and depressions on the under side of the scuta, which are in direct connexion with the muscles and soft parts of the animal, vary to a certain extent: thus the length and prominence of the adductor ridge is decidedly variable in the fossil *B. concavus* and *tintinnabulum;* the size and form of the little cavity for the lateral depressor muscle varies in many species; so does the exact shape and degree of prominence of the articular ridge. There is one character in the terga, which at first would be thought very useful, namely, whether an open longitudinal furrow, or a closed fissure runs down outside the valve from the apex to the spur; but it is found that the furrow almost always gradually closes up during growth; and as a consequence of this, the width of the spur compared to that of the whole valve, as well as its distance from the basi-scutal angle, and the form of its basal extremity, all vary in some degree. The length of the spur sometimes varies considerably, as in *B. concavus.* The summits of the radii are apt to be oblique in the young of some species, whereas they are generally quite square in the old of the same. In some species the scuta become longitudinally striated only with age; on the other hand, in very young specimens of *B. tintinnabulum*, the scuta sometimes are deeply impressed by little pits placed in rows. I have already alluded to the longitudinal furrow on the tergum so entirely changing its character, owing to the edges becoming, during growth, folded inwards, and to the consequences which result from this. The inner lamina of the parietes generally loses, to a certain extent, its longitudinally ribbed character in old age. The basis is solid instead of being porose, in very young specimens of some species. In all the species, the carino-lateral compartments, in early age, are very narrow in proportion to the width of the lateral compartments; and in all, at this early period, the operculum is large in proportion to the whole shell.

Finally, I must state my deliberate conviction that Sessile Cirripedes can very seldom be satisfactorily identified in a fossil condition, without an examination of the opercular valves: hence when these have not been discovered, I have resolved, with some rare exceptions, not to attach a specific name to a shell without its operculum; for thus, I believe, I should add to the number of useless synonyms, which, as we shall immediately see, already exist. Nothing, indeed, could have been easier than to have affixed names to many groups of specimens, having different aspects, but to feel sure that these were really distinct species requires better evidence than can be afforded by the shell, without the opercular valves. When the specimens are much fossilised, it is, indeed, difficult to make out the primary points of structure in the genus Balanus—namely, whether the parietes, radii, and basis are porose: to do this it is sometimes necessary to rub down, polish, and carefully examine, a transverse section of a piece of the shell.

The ancient history of the Balanidæ is a brief one. No Secondary species has hitherto been discovered; in my former monograph on the fossil Lepadidæ[1] I have shown that the negative evidence in this case is of considerable value, and consequently that there is much reason to doubt whether any member of the family did exist before the eocene period.

[1] Since the note to page 5 of that work was written, I have been informed that the so-called cretaceous *Tubicinella maxima* is not a Cirripede.

The existence of a Cretaceous Verruca is an apparent exception to the rule, as this genus has hitherto always been ranked among Sessile Cirripedes; but Verruca, as we now know, must be placed in a family by itself, namely, the Verrucidæ, quite distinct from the Balanidæ. Balanus is the oldest genus as yet known; it first appeared in Europe and North America, during the deposition of the eocene beds; and was at that time, as far as our information at present serves, represented by very few species. In South America, one species of Balanus abounds in individuals in the ancient Patagonian tertiary formation. I have seen, in the British Museum, specimens said to have come from the eocene nummulitic beds, near the mouth of the Indus, belonging to that section of the genus, which has the walls and basis permeated by pores. Generally, the extinct forms belong to the section, which has the parietes not permeated by pores. During the miocene and pliocene ages, Sessile Cirripedes abounded. No extinct genus in this family has hitherto been discovered. It is singular, that though the Chthamalinæ approach much more closely than do the Balaninæ to the ancient Lepadidæ, of which so many species have been found fossil even in the older Secondary formations, yet that only one species of one genus of this sub-family has been hitherto found in any deposit; and that species is the still existing *Pachylasma giganteum*, in the modern beds of Sicily. During the epoch of the Glacial deposits in Scandinavia, Scotland, and Canada, the still existing species seem to have abounded; and they attained larger average dimensions than the same species now do on the shores of Great Britain, or even on the shores of the northern United States, where the average size seems larger than on this side of the Atlantic.

I already have given my reasons for very seldom naming any Sessile Cirripede without examining the opercular valves: it has been owing to this, as it appears to me, proper want of caution, that there are so many nominal species. Thus it is made to appear in catalogues, that the tertiary seas abounded with species of Balanus to an extent now quite unparalleled in any quarter of the world. Bronn,[1] for instance, in his invaluable 'Index Palæontologicus,' gives the names of thirty-five Balani, found fossil in Europe, and I have not counted those found only in alluvial deposits, as they would certainly be the same as the still living species. Now I know only eleven recent Balani on the shores of all Europe, from the North Pole to lat. 30°; and of these I doubt whether *B. balanoides* and *improvisus* have been found fossil. In the Red Crag there is one extinct Balanus: in the Coralline Crag, which seems to have been very favorable to the existence of Cirripedes, there are six species of Balani, of which two are absolutely extinct, and one does not occur in any neighbouring sea: in the Eocene formations the species seem to have been rare, and I have seen only one, and that is an extinct form. Taking these several facts into consideration, and bearing in mind that Cirripedes usually range widely,

[1] To save any other person interested in fossil Cirripedia, going through the several works quoted by Bronn, I have given some remarks on his list of species, in an appendix at the end of the Balanidæ, in my volume published by the Ray Society.

I do not believe, if all the specimens of Balani hitherto found in the several tertiary formations, from the Eocene to the Glacial deposits, throughout Europe, were collected together, they would amount to twenty species. I have myself seen, in a recognisable state, only twelve fossil species, of which five are extinct or not found in any neighbouring sea: I think it probable that three other recent species, viz., *B. tulipiformis, perforatus,* and *amphitrite,* may occur in the Mediterranean formations: and this would make fifteen species. Therefore, in the above estimate of twenty species, five are allowed for species existing in European collections, but not hitherto seen by me; and this, I believe, is a very full allowance. Consequently, even on the supposition that the five species just admitted as possibly existing in cabinets, and that the other five extinct species, which I have seen and examined, have all been previously named by other authors, a supposition excessively improbable, even then there would be fifteen superfluous names in Bronn.

The following short table shows how the Balanidæ and Verrucidæ were represented in Great Britain, throughout the several TERTIARY STAGES. It includes all the sixteen species described in the following pages, with the exception of one, the cretaceous *Verruca prisca,* which is the only member of either family hitherto found in any Secondary deposit.

Name.	Living species, but found fossil in some tertiary deposits.	Mammaliferous crag, and glacial deposits.	Red crag.	Coralline crag.	Eocene.
Balanus tintinnabulum	*	..	*	..	..
calceolus	*	..	..	*	..
spongicola	*	..	..	*	..
concavus	*	..	*	*	..
porcatus	*	*	*	..	..
crenatus	*	*	*	*	..
Hameri	*	*	*	..	..
bisulcatus	..	..	..	*	..
dolosus	..	*	*	..	..
inclusus	..	..	..	*	..
unguiformis	..	..	..	..	*
Acasta undulata	..	..	..	*	..
Pyrgoma anglicum	*	..	..	*	..
Coronula barbara	..	..	*	..	..
Verruca Strömia	*	*	*	*	..
Total 15, recent and extinct, found fossil in Great Britain, in some tertiary deposit	9	5	8	9	1

As affording some standard of comparison by which to compare the number of species found fossil in any Tertiary deposit, in relation to the number of species probably existing in the neighbouring seas during the same epoch, I may state that there are now living and propagating on the shores of Great Britain, eleven species belonging to the two Families included in the above table. In the Coralline crag, which seems to have been eminently favorable for the existence and subsequent preservation of Cirripedes, and which has been so well worked, nine fossil species of these two families, as may be seen in the table, have been discovered.

NOMENCLATURE OF THE SHELL OF A SESSILE CIRRIPEDE.

ARCHETYPE SHELL. Fig. 1.

Orifice of shell, surrounded by the *sheath. Sheath* formed by the *alæ* (*a—a*) and by portions of the upper and inner surfaces of the *parietes* (*p—p*).

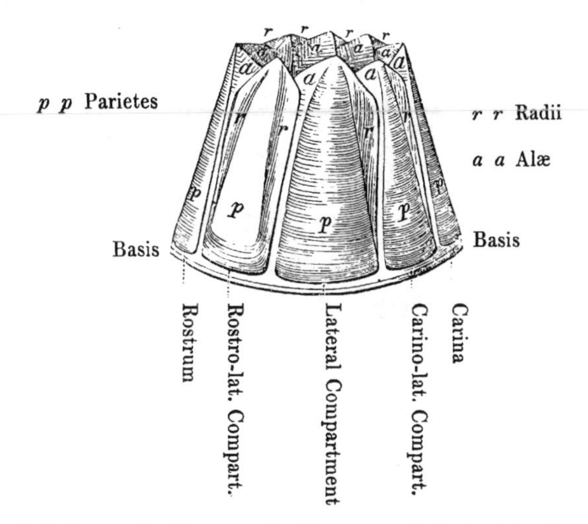

N. B. In Balanus, and all known fossil genera, the Rostrum and Rostro-lateral compartments are confluent, and hence the Rostrum has the structure of Fig. 2.

COMPARTMENTS.

Fig. 2.

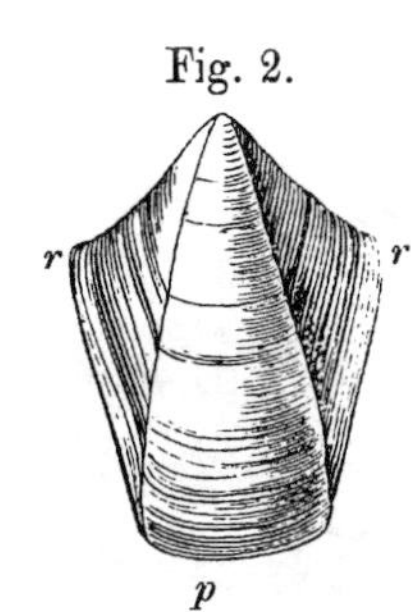

Fig. 3.

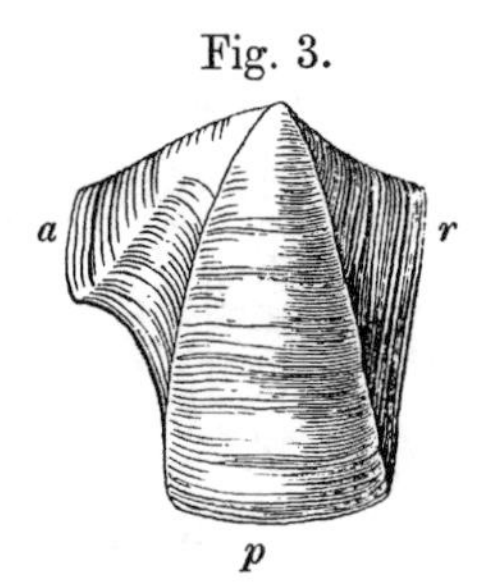

Fig. 4.

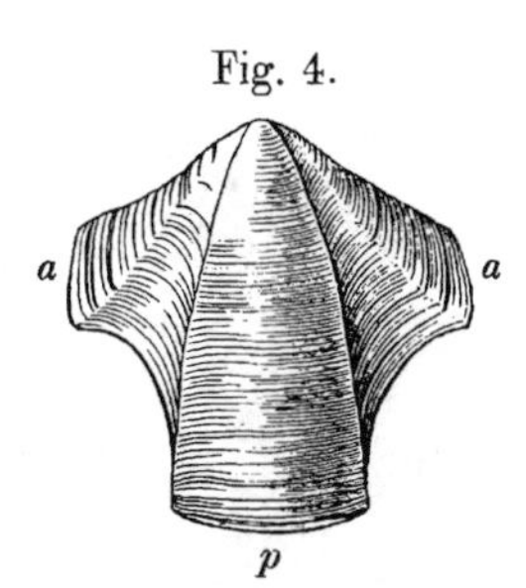

Fig. 2. Compartment with two radii, serving, in fossil specimens, always as a Rostrum.
Fig. 3 serves as a Lateral and Carino-lateral Compartment. Fig. 4 serves as a Carina.

OPERCULAR VALVES.

Fig. 5. Scutum (internal view).

Tergal Margin
Occludent Margin
Articular Ridge
Cavity for Adductor Muscle
Articular Furrow
Adductor Ridge
Cavity for the Lat. Depressor Muscle.

Fig. 6. Tergum (internal view).

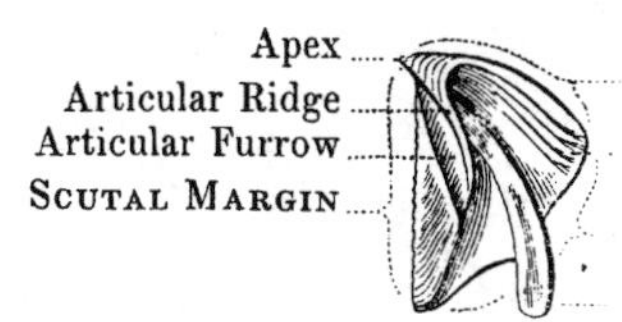

Fig. 7. Tergum (external view).

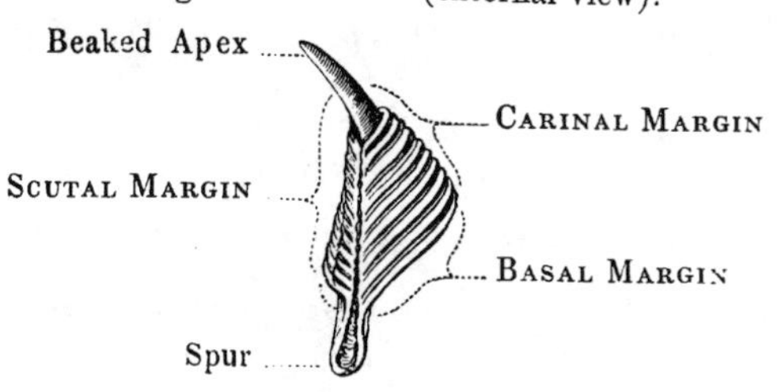

On the Names given to the different parts of Cirripedes.

In my former volume I stated that I had found it indispensable, in part owing to the extreme confusion of the nomenclature previously used, to attach new names to several of the external parts of Cirripedes. Almost all these names are applicable to the Balanidæ, or Sessile Cirripedes, and to the Verrucidæ; but a few additional names are requisite, which, together with the old names, will, I hope, be rendered clear by the accompanying woodcuts. In Sessile Cirripedes, the whole of that which is externally visible, may, for convenience sake, be divided into the *operculum* or *opercular valves* (*valvæ operculares*), and the *shell* (*testa*), though these parts homologically present no real difference. The operculum is seated generally some little way down within the *orifice* of the shell; but in very young specimens, and in Verruca, the operculum is attached to the summit of the shell, and in these cases the shell, without the operculum be removed, can hardly be said to have any orifice; though, of course, the opercular valves themselves have an aperture for the protrusion of the cirri.

The shell consists of the *basis* (called the support by some authors), and of the *compartments* (*testæ valvæ*), which in recent specimens vary from eight to four in number, and occasionally are all calcified together.

The compartment, at that end of the shell (fig. 1) where the cirri are exserted through the aperture or lips of the operculum, is called the *carina* (fig. 4); the compartment opposite to it, is the *rostrum* (in all fossil specimens, like fig. 2),—these two lying at the ends of the longitudinal axis of the shell. Those on the sides are the *lateral compartments;* that nearest the carina, being the *carino-lateral* (fig. 3) (*testæ valva carino-lateralis*), that nearest the rostrum, the *rostro-lateral,* and middle one simply the *lateral* compartment (fig. 3): but these three compartments are rarely present together. The *rostro-lateral* compartment, which always resembles fig. 2, and may be always known by having radii on both sides, is not known to occur in any fossil species; and hence we are here only concerned with the lateral and carino-lateral compartments. The compartments are separated from each other by *sutures,* which are often so fine and close as to be distinguished with difficulty. The edge of a compartment, which can only be seen when disarticulated from its neighbour, I have called the sutural edge (*acies suturalis*).

Each separate *compartment* consists of a *wall* (*paries*), or *parietal portion* (*pp* in figs. 1 and 4), which always grows downwards, and forms the basal margin; and is furnished on the two sides either with *alæ* (fig. 4), or with *radii* (fig. 2), or with an ala on one side and a radius (fig. 3) on the other.

The *radius* (adopting the name used by Bruguière, Lamarck, and others) differs remarkably in appearance (though not in essence) from the wall or parietal portion, owing to the direction of the lines of growth and the state of its usually depressed surface. In the upper part, the radii overlie the alæ of the adjoining compartments: in outline (*r*, fig. 2, 3), they are wedge-formed, with their points downwards; their summits (and this is often a useful specific character) are either parallel to the basis, or as in fig. 1, 2, oblique. The radii are sometimes not developed.[1]

The *alæ* (so called by Dr. Gray) are overlapped by the radii, and by part of the walls; they usually extend only about half way down the compartment (*a*, fig. 3, and 4); their summits are either parallel to the basis or oblique. The alæ of the several compartments, together with the internal, upper, thickened surfaces of the walls, against a shoulder of which the sutural edges of the alæ abut, have been called (by Dr. Gray) the *sheath* (*vagina*). The upper and greater portion of the sheath is marked by transverse lines, caused by the exuviation of the *opercular membrane*, as that membrane may be called, which unites the operculum all round to the sheath, or upper internal surface of the shell.

The *carina* has always two *alæ*, as in fig. 4.

The *carino-lateral* and *lateral compartments* have always an *ala* on one (the rostral) side, and a *radius* on the other (the carinal) side, as in fig. 3.

The *rostro-lateral compartment* (not at present known to occur in any fossil) has always *radii* on both sides, as in fig. 2.

The *rostrum* has normally *alæ* on both sides, as in fig. 4; but in many recent, and all the fossil species yet known, it has *radii* on both sides, as in fig. 2, owing to its fusion with the rostro-lateral compartments on both sides.

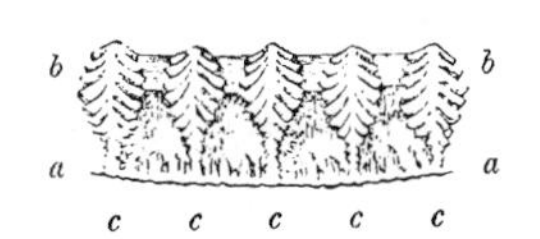

Basal edge of wall of compartment in *Balanus tintinnabulum*. *a*, *a*, outer lamina; *b*, *b*, inner lamina; *c*, *c*, longitudinal septa, uniting the inner and outer laminæ, with their ends denticulated.

The walls of the shell, the basis, and the radii, are in very many cases composed of an *outer* and *inner lamina*, united together by *septa;* a set of pores or tubes being thus formed. The points of the septa generally project beyond the laminæ, and are denticulated on both sides, as shown in the accompanying woodcut.

Operculum, or *opercular valves*.—These consist of a pair of scuta and a pair of terga. They are joined to the sheath of the shell by the *opercular membrane*.

Scutum (fig. 5): this valve is generally sub-triangular, and its three margins are the *basal*, the *tergal*, so called from being articulated with the tergum, and the *occludent*, so called from opening and shutting against the opposed valve. The angles are named, from the adjoining margins, as *basi-tergal*, &c.; the upper angle being the apex. The scutum is ordinarily articulated to the tergum by an *articular ridge* (*crista articularis*), running up to the apex of the valve, and by an *articular furrow*, which latter receives the

[1] The radii have been called by Ranzani and De Blainville "areæ depressæ" (the parietal portions of the compartments being the "areæ prominentes"); by Poli, "areæ interjectæ;" by Gray, "sutures;" by Coldstream, "compartments of the second order," (the parietal portions being those of the first order); by some authors, "intersticia." I may here add that the scuta are the "ventral valves" of Gray, the "anterior" of Ranzani, and the "inferior opercular" of De Blainville: the terga are the "posterior valves" of Gray and Ranzani, but the "superior opercular" of De Blainville: the rostrum, on the other hand, is the "anterior valve" of Ferussac, and the "ventral" of De Blainville; the carina being the "dorsal valve" of the latter author.

scutal margin of the tergum. The articular ridge, instead of projecting straight up from the valve, when the latter is laid flat on its external surface, often bends over to the tergal side, and is then said to be *reflexed*. On the internal surface of the valve there is almost always an *adductor pit* or *cavity* (*fossa adductoris*), for the attachment of the adductor scutorum muscle: this pit is often bounded on its tergal and basal sides, by a ridge, called the *adductor ridge* (*crista adductoris*), which, in its upper part, is often confluent with the articular ridge. Beneath the adductor ridge, in the basi-tergal corner of the valve, there is often a *lateral-depressor pit* (*fossa musculi lateralis depressoris*), for the attachment of the so-called muscle; and this pit is sometimes furnished with crests.

Tergum, (figs. 6 and 7):—this valve, also, has three margins, the *scutal*, *basal*, and *carinal*; its upper end, or *apex*, is sometimes *beaked*; on the basal margin a *spur* (*calcar*) depends; the outer surface of the valve is depressed or longitudinally *furrowed* (*sulcus longitudinalis*) in the line of the spur. The part called the spur is often so broad, that the name becomes not very appropriate. The angles are denominated, from the adjoining margins, as *basi-carinal*, or *basi-scutal* angle, &c. On the under side, in the upper part, there is an *articular ridge*, and on its scutal side, an *articular furrow*, receiving the articular ridge of the scutum. In the basi-carinal corner of the valve there are often crests for the attachment of the tergal depressor muscle.

Relative position of parts.—The centre of the generally flat basis, which is cemented to the supporting surface, is properly the *anterior* end, and the tips of the terga, often hidden within the shell, are properly the *posterior* end of the external covering; but I have found it more convenient to speak of the *upper* and *basal* surfaces and aspects, which hardly admit of any mistake. A line drawn from the centre of the basis, along the middle of the rostrum to the tips of the scula, shows the strictly *medio-ventral* surface of the shell; and another line drawn from the centre of the basis, along the carina, to the tips of the terga, shows the strictly *medio-dorsal* line; but from the crooked course of these lines, I have found it far more convenient to speak of the *rostral* and *carinal* end or aspect of the different parts of the shell. There has, moreover, been great confusion in these relative terms, as applied by different authors.

When a sessile Cirripede is held in the position in which they have generally been figured, namely with the basis downwards and the scuta towards the beholder, then the *right* and *left* sides of the Cirripede correspond with those of the holder.

SUB-CLASS—CIRRIPEDIA. ORDER—THORACICA.

Family—BALANIDÆ.

Cirripedia sine pedunculo; scuta et terga musculis depressoribus instructa; reliquæ testæ valvæ inter se immobiliter conjunctæ.

Cirripedia without a peduncle; scuta and terga furnished with depressor muscles; other valves united immoveably together.

This family, which includes all true Sessile Cirripedes, may be divided into two very natural sub-families; namely, the Balaninæ and Chthamalinæ; but as not one member of the latter has been found fossil in Great Britain, and indeed only one, the *Pachylasma giganteum,* in any part of the world, viz., in the recent beds of Sicily, this sub-family of the Chthamalinæ may be here passed over in silence.

Sub-Family—BALANINÆ.

Rostrum cum radiis, sed sine alis: valvæ testæ laterales omnes, ex uno latere alis, ex altero radiis instructæ: parietes ferè aut porosi, aut ad interiorem superficiem longitudinaliter costati.

Shell with the rostrum having radii, but without alæ; lateral compartments all having alæ on one side and radii on the other side; parietes generally either porose, or longitudinally ribbed on their inner surfaces.

Genus—BALANUS, *Auct.*

CONOPEA (pars generis). *Say.* Journal Nat. Sc. Philadelphia, vol. ii, Part 2, 1822.
MESSULA (do.) *Leach.* Zoological Journal, vol. ii, 1825.
CHIRONA (do.) *J. E. Gray.* Philosoph. Transacts., 1835, p. 37.

Valvæ operculares inter se articulatæ, subtriangulares; valvæ testæ 6*; basis calcarea aut membranacea.*

Scutum and tergum articulated together, sub-triangular; compartments six; basis calcareous or membranous.

The genus Balanus already includes 45 species, recent and fossil, and consequently in my volume published by the Ray Society, I have divided the genus into sections, on characters derived from the porosity of the parietes, radii, and basis; and on whether the basis be membranous or calcareous. But as here we have to describe or notice only eleven species, I have thought it more convenient to drop the sections, and in their place add a few words to each of the diagnostic characters. The genus is quite distinct from all the other genera of Sessile Cirripedes, with the exception of the sub-genus Acasta, from which its separation, it must be confessed, is in one sense artificial; for the species of this sub-genus graduate into those of Balanus (such as *B. calceolus* and its allies), which have their shells elongated in the rostro-carinal axis, and which live attached to Gorgoniæ. These latter species have been generically separated by some authors from true Balanus; but I have found it impossible to effect this; and even the section of the genus, including these species, is hardly distinct enough from the adjoining sections. On the other hand, the sub-genus Acasta, in another sense, is a very natural one, inasmuch as all its species are closely allied together in essential structure, in general appearance, and in habit; and as the genus Balanus is already large, I have thought it best to adopt Acasta, which has been already admitted by many authors as a sub-genus. I need only further remark, that from reasons already assigned, I have thought it useless to give in this work long generic descriptions.

1. Balanus tintinnabulum. Tab. I, fig. 1*a*—1*d*.

Lepas tintinnabulum. *Linn.* Syst. Naturæ, 1767.
— — *Ellis.* Phil. Transact, vol. 50, 1758, Tab. 34, figs. 8 and 9.
— — *Chemnitz.* Neues. Syst. Conch., 8 B. (1785), Tab. 97, figs. 828–831.
Balanus tulipa. *Bruguière.* Encyclop. Meth., 1789; sed non *B. tulipa alba,* in *Chemnitz;* nec non *B. tulipa, O. F. Müller,* Zoolog. Dan.; nec non *B. tulipa,* Poli, Test. ut Siciliæ.
— — *G. B. Sowerby.* Genera of Recent and Fossil Shells, Tab. Genus Balanus.
Lepas crispata (*var.*) *Schröter.* Einleitung Conch., vol. iii, Tab. 9, fig. 21.
— spinosa (*var.*) *Gmelin.* Linn. Syst. Nat.
— tintinnabulum, spinosa, crispata et porcata. *W. Wood.* General Conchology, 1815, Pl. 6, figs. 1, 2. Pl. 7, figs. 4, 5. Pl. 8, figs. 1—5.
Balanus tintinnabulum. *Chenu.* Illust. Conch.
— d'Orbignii (*var.*) *Chenu.* Illust. Conch., Tab. 6, fig. 10, sed non Tab. 4, fig. 13.
— crassus. *Sowerby* (!) Min. Conch., 1818, Tab. 84.

B. parietibus et basi et radiis poris perforatis: testâ a roseâ ad atro-purpuream variante, sæpe longitudinaliter virgatâ et costata: orificio plerumque integro, interdum dentato. Scuti cristâ articulari latâ et reflexâ. Tergi margine basali plerumque in contrariis calcaris partibus rectam lineam formante.

Walls, basis, and radii permeated by pores; shell varying from pink to blackish purple, often striped and ribbed longitudinally; orifice generally entire, sometimes toothed. Scutum with the articular ridge broad and reflexed. Tergum with the basal margin generally forming a straight line on opposite sides of the spur.

Fossil in the Red Crag (Sutton). Mus. S. Wood, J. de C. Sowerby. Touraine (?) Mus. Lyell.

Recent, on West Coast of Africa; Madeira; West Indies; Cape of Good Hope; Mouth of the Indus; East Indian Archipelago; Sydney, Australia; Peru; Galapagos Islands; West Mexico; California.

Of this species I have seen several specimens, and fragments. Three of these are the original specimens figured in the Mineral Conchology, as *B. crassus*, an examination of which I owe to the great kindness of Mr. J. de C. Sowerby. Some specimens equally or more perfect are in Mr. S. Wood's collection. I have further seen a specimen from Touraine, which was presented to Sir C. Lyell by M. Dujardin, under the name of *B. fasciatus*, which I fully believe to be *B. tintinnabulum*. None of these specimens had opercular valves, and therefore it is perhaps rash to assert quite positively that they are identical with *B. tintinnabulum;* but, extraordinarily variable as this latter species is, yet, after having examined so many hundreds of recent specimens from all quarters of the globe, a sort of instinctive knowledge of general aspect is acquired, which makes me feel convinced that the fossils in question do really belong to this species. Moreover, the large shell, with its trigonal orifice passing into rhomboidal,—the smooth, broad, finely porose radii, with their summits not oblique,—the rather large parietal pores,—and the cancellated basis, are characters which hardly concur in any other species; and those with which these fossils might be confounded, are inhabitants of distant quarters of the world. Most of the recent varieties of *B. tintinnabulum*, and all the fossil specimens from the Crag, can be at once discriminated from *B. tulipiformis* (with which, at least in the Mediterranean deposits, it is likely to be confounded) by the summits of the radii extending from tip to tip of the adjoining compartments, and therefore not being oblique, as is always the case with the radii of *B. tulipiformis*. The largest fossil specimen which I have seen is nearly two inches in basal diameter, and nearly the same in height, and therefore about two thirds of the size of the largest living specimens.

I have had engraved, from recent specimens, an internal view of the scutum and tergum, as these are likely hereafter to be found by searchers in the Crag deposits; and I may refer to my Monograph on the Balanidæ for their full description. It may be observed in the habitats given of the living specimens, that Madeira is the nearest point where the species now lives and propagates; but specimens in full vigour are often brought to the British shores, attached to the bottoms of vessels.

2. Balanus calceolus,* Tab. I, fig. 2*a*—2*d*.

Balanus calceolus keratophyto involutus (?) *Ellis*. Phil. Trans., vol. 50 (1758), Tab. 34, fig. 19.
Lepas calceolus (?) *Pallas*. Elench. Zooph., p. 198, (sine descript.) (1766).
Conopea ovata (?) *J. E. Gray*. Annals of Philosophy, vol. 10, 1825.

B. testæ axe rostro-carinali elongato; basi cymbiformi; parietibus et basi, sed non radiis, poris perforatis. Scuto musculi depressoris lateralis fossâ parvâ, profundâ.

Shell with its rostro-carinal axis much elongated; basis boat-shaped; walls and basis porose, but not the radii. Scutum with the pit for the lateral depressor muscle small and deep.

Fossil in Coralline Crag, attached to a Gorgonia; Sutton; Mus. S. Wood.
Recent, attached to Gorgoniæ, West Coast of Africa. Tubicoreen, near Madras. Mediterranean (?).

I have seen only a single fossil specimen of this species, nearly half an inch in length. The shell was perfect, and a small portion of the Gorgonia yet remained attached to the grooved and boat-shaped basis. The opercular valves had been lost, but the shell in this instance is so peculiar, that it could only be confounded with the recent *B. galeatus, cymbiformis,* or *navicula,* and from all these it is easily distinguished by the parietes being permeated by pores. It is, of course, possible, that the opercular valves might present some new character, showing that this fossil, though agreeing with *Bal. calceolus* in its shell, yet was specifically distinct. I have given a drawing of the opercular valves from recent specimens, which have been fully described in my Monograph on the Balanidæ. In regard to the shell, the fossil specimen could not be distinguished from the recent; and as it had to be broken, in order that its internal structure might be examined, I have thought it best to give a drawing from a perfect recent specimen. The spur of the tergum, in recent specimens, sometimes presents a singular character, in being irregularly toothed, and I have given a drawing (fig. 2d) of this variety, as it might perplex a collector.

[1] With respect to the nomenclature of this and three allied recent species, I must remark that in the published descriptions no allusion is made to any one of the characters by which alone they can be distinguished: hence I have been guided by geographical probabilities in assigning the specific name of *calceolus* to the present species, as Ellis's specimens came from the Mediterranean; and that of *galeatus* to the North American and West Indian specimens, as Linnæus' original specimens (according to a statement by Spengler) came from the West Indies. I have assigned new names to the two remaining East Indian species.

3. Balanus spongicola, Tab. I, fig. 3*a*—3*e*.

Balanus spongicola. *Brown's* Illustrations of the Conchology of Great Britain (1827), pl. 7, fig. 6: 2d edit. (1844), pl. 53, figs. 14—16.

B. parietibus et basi, sed non radiis poris perforatis; parietibus plerumque lævibus, roseis; orificio dentato; scuto longitudinaliter striato; tergum, apice producto, sine sulco longitudinali, calcare truncato, $\frac{1}{3}$ *valvæ latitudine.*

Parietes and basis, but not the radii, permeated by pores; parietes generally smooth; shell pink; orifice toothed; scutum longitudinally striated; tergum, with the apex produced, without a longitudinal furrow; spur truncated, about one third of width of valve.

Fossil in Coralline Crag; Sutton; Mus. S. Wood.

Recent on the South coast of England, and Tenby in South Wales; Algiers; Madeira; Lagulhas Bank, Cape of Good Hope.

I have seen only a single specimen of this species, which I picked out of a mass of specimens of the extinct *Bal. inclusus*, collected by Mr. Wood, in the Coralline Crag at Sutton. This one specimen was perfect, and included the opercular valves; it even partially retained its rosy colour: it was ·3 of an inch in basal diameter, and therefore exactly half the size of the largest recent specimen which I have seen. It was in every respect perfectly characterised. I have given drawings, external and internal, of the scutum and tergum from the fossil specimens. In the scutum, the adductor ridge is, perhaps, rather more prominent, and the pit for the lateral depressor muscle rather deeper than in recent specimens; but these points are extremely variable. The tergum, in its outline, strictly agrees with the European recent specimens, and not with those varieties from the Cape of Good Hope and West Indies; indeed, in the degree in which the basal margin on the carinal side of the spur slopes towards the spur, it even, perhaps, exceeds the European variety. These valves are fully described in my Monograph on the Balanidæ. From the shell alone, as viewed externally, *Bal. spongicola,* even in its recent state, can hardly be distinguished from *Bal. tulipiformis,* or from some varieties of *Bal. Capensis:* I doubt whether this species could anyhow be distinguished in its fossil condition from the young of the fossil *Bal. concavus,* without the aid of the opercular valves. But in order to give an idea of its general appearance, and as I was compelled to disarticulate the compartments of the one fossil shell, I have had a fine recent specimen from the Mediterranean engraved on an enlarged scale.

4. Balanus concavus, Tab. I, fig. 4*a*—4*p*.

Balanus concavus. *Bronn.* Italiens Tertiar-Gebilde (1831) et Lethæa Geognostica, b. ii, s.1155 (1838), Tab. 36, fig. 12.[1]
— cylindraceus, *var.* c. *Lamarck.* Animaux sans Vertèbres (1818).
Lepas tintinnabulum. *Brocchi.* Conchologia Sub-Appen., t. ii, p. 597, (1814).

B. parietibus et basi, sed non radiis poris perforatis; testâ albo cum roseo aut obscurè purpureo longitudinaliter pictâ, interdum purè albâ. Scuto longitudinaliter tenuiter striato: internè, adductoris cristâ admodum aut modicè prominente.

Parietes and basis, but not the radii, permeated by pores; shell longitudinally striped with white and pink, or dull purple; sometimes wholly white; scutum finely striated longitudinally; internally, adductor ridge very or moderately prominent.

Fossil in Coralline Crag, (Ramsholt and Sudbourne) rarely in the Red Crag (Sutton); Mus. S. Wood, Bowerbank, Lyell, J. de C. Sowerby, Tennant. Sub-Appenine formations, near Turin, Asti, and Colle in Tuscany, Mus. Greenough, &c. Tertiary bed, near Lisbon, Mus. D. Sharpe and Smith. Bordeaux (?) Mus. Lyell. Tertiary beds, Williamsburg; and Evergreen, Virginia, Mus. Lyell. Maryland, Mus. Krantz. Pleiocene formations[2] near Callao, Peru, Mus. Darwin.

Recent at Panama; Peru; S. Pedro, California; Philippine Arch.; Australia.

This species has caused me much trouble. It will be convenient first to make a few remarks on the recent specimens; I examined several from Panama and California, which, though differing greatly in colour, resembled each other in their scuta having the adductor ridge extremely prominent, and in having (Tab. I, fig. 4*n*) an almost tubular cavity for the attachment of the lateral depressor muscle,—characters which at first appeared of high specific value; but I soon found other specimens from Panama in which these peculiarities were barely developed. I then examined a single specimen from the Philippine Archipelago, resembling in external appearance one of the Panama varieties, but differing in the scuta being externally strongly denticulated in lines instead of being merely striated,—in the adductor ridge being far less prominent,—and in the spur of the tergum being broader and more truncated; I therefore considered this as a distinct species. I then examined a single white rugged specimen from the coast of Peru, which differed from the Philippine specimen in the shape of the well-defined denticulations on the scuta, and in some other trifling respects, and in the segments of the posterior cirri bearing a greater number of spines; with considerable doubt, I also named this as distinct. But when I came to

[1] I suspect that *B. pustularis, miser,* and *zonarius,* all figured by Münster, in his 'Beiträge,' b. iii, Tab. 6, may be this species.

[2] I procured this specimen from the Island of S. Lorenzo, off Callao; it was imbedded, together with seventeen species of recent shells and with human remains, at the height of eighty-five feet.

examine a large series of fossil specimens from the Coralline Crag of England, and others from northern Italy, from Portugal, and from the southern United States, I at once discovered that the form of the denticuli on the striæ of the scuta was quite a worthless character,—that in young specimens the scuta were simply striated,—that the prominence of the adductor scutorum ridge and the depth of the cavity for the lateral depressor muscle varied much (as in the case of the recent specimens), owing apparently to the varying thickness of the valve,—that in the terga the spur varied considerably in length and breadth, the latter character being in part determined by the varying extent to which the edges of the longitudinal furrow are folded in,—and lastly, that in young specimens the basal end of the spur is much more abruptly truncated than in the old. Hence I was led to throw the three recent forms, originally considered by me as specifically distinct, into one species; but I may repeat that this considerable variation in the prominence of the adductor ridge, and in the depth of the pit for the lateral depressor muscle—the pit in some cases becoming even tubular—is a very unusual circumstance.

With respect to the fossil specimens from the above stated distant localities, I consider them as belonging to one species, though they vary considerably in several points of structure. When compared with the recent specimens, they differ from them in often attaining a considerably larger size; in the parietes being generally longitudinally ribbed, as in the case of the Coralline Crag specimen (Tab. I, fig. 4*a*), and in the radii often having more oblique summits. Some of the specimens from the United States, have strong rugged, depressed shells, frequently resembling, to a curious degree, *Bal. porcatus.* On the other hand, considering the many points of identity between the fossil and the recent specimens, I have concluded, without much doubt, that they ought all to be classed together. In the Coralline Crag specimens, the spur of the tergum (Tab. I, fig. 4*g*) is unusually long and narrow; it is broader in the Italian specimens (4*o*), and either short (4*k*) or long in the United States specimens. The scuta of the Lisbon specimens are remarkable for the great prominence of the adductor ridge, and for the depth of the lateral depressor cavity, as in most, but not in all, of the Panama specimens. The opercular valves, however, of some of the specimens from all these several distant localities are identical with the recent ones from the coast of America. I have entered into the above particulars, on account of, in the first place, its offering an excellent example how hopeless it is in most cases to make out the species of this difficult genus without a large series of specimens; secondly, as showing how the characters alter with age; and thirdly, as a good instance of the amount of variation which seems especially to occur in most of the species which have very extensive ranges.

Some of the pink-striped Panama varieties, though having a somewhat different aspect, can be distinguished from certain varieties of *B. amphitrite* only by their scuta being longitudinally striated,—a character in this species variable in degree, and in most cases of very little value. Some of the other recent varieties, however, are sufficiently distinct from *B. amphitrite;* and the great fossil Coralline Crag specimens, which stand at the opposite end of the series of varieties, with their ribbed walls, very oblique radii, and coarsely striated scuta, are extremely unlike *B. amphitrite.*

With respect to the nomenclature of the present species, I have little doubt that I have properly identified the Italian fossil specimens with *B. concavus* of Bronn, who has given a very good figure of this species in his 'Lethæa Geognostica;' but it must be confessed that the longitudinal striæ on the scuta are not there represented. Considering the large size and frequency of this species in Europe and in the United States, it has probably received several other names besides the two synonyms, quoted at the head of this description. I should add that the true *B. cylindraceus* (not *var.* c) of Lamarck, according to the plate given by Chenu in his 'Illust. Conch.' is the *B. psittacus* of South America. I have seen in collections specimens of *B. concavus* labelled as *B. tulipa* of Poli (*B. tulipiformis* of my Monograph),—a very natural mistake, without the opercular valves be carefully examined.

General Appearance.—Shell conical (fig. 4*a*), often steeply conical (fig. 4*c*), but sometimes depressed and smooth (fig. 4*d*); orifice generally rather small, varying from rhomboidal to trigonal, with the radii narrow, and generally in the fossil specimens very oblique; surface generally smooth, sometimes rugged, and in the Coralline Crag specimens commonly ribbed longitudinally, the ribs being narrow. In the recent specimens the colour is various, either dull reddish-purple with narrow nearly white, or wider dark longitudinal bands; or, again, pale rosy-pink with broad white bands; or lastly, wholly white. The radii are either darker or paler than the parietes. The opercular valves are either dark purple or nearly white. Pale pink and white stripes are visible on some of the Italian and Portuguese tertiary specimens; and in most of the fossils the sheath is tinged dull red.

Dimensions.—The largest actually recent specimen which I have seen, from the Philippine Archipelago, had a basal diameter of 1·2 of an inch; the Peruvian pleistocene specimen is 1·7 in diameter; specimens from the crag and from the Italian deposits, however, sometimes slightly exceed two inches in basal diameter, and three in height.

Scuta: these in young and moderately-sized specimens are striated longitudinally (fig. 4*l*), sometimes faintly, but generally plainly, causing the lines of growth to be beaded; but in large and half-grown specimens, the lines of growth are often extremely prominent, and being intersected by the radiating striæ, are converted into little teeth or denticuli. As the striæ often run in pairs, the little teeth frequently stand in pairs, or broader teeth have a little notch on their summits, bearing a minute tuft of spines. In very old and large specimens, the prominent lines of growth are generally simply intersected by deep and narrow radiating striæ (tab. I, fig. 4*p*). In one case, a single zone of growth in one valve was quite smooth, whilst the zones above and below were denticulated. The valve varies in thickness, which I think influences the prominence of the lines of growth and the depth of the striæ. These striæ often affect the internal surface (fig. 4*h*) of the basal margin, making it bluntly toothed. The articular ridge (fig. 4*n*), is rather small, and moderately reflexed. The adductor ridge (as already stated) varies remarkably; in most of the recent Panama specimens (fig. 4*n*), and in the fossils from Portugal, it is extremely prominent, and extends down to near the basal margin; in other specimens it is but slightly prominent, as in those from the Crag (4*f*); it is short, but rather prominent in the specimens (4*h*) from Maryland; whereas it is very slightly prominent in the specimens from Virginia. The cavity for the lateral depressor, also, varies greatly; it is often, as in the recent specimens, bounded on the side towards the occludent margin by a very slight straight ridge, which occasionally folds a little over, making almost a tube; this, at first, I thought an excellent specific character, but far from this being the case, the cavity often becomes, in recent specimens as well as in the crag specimens (4*f*), wide, quite open, and shallow. The whole valve in the Crag specimens (fig. 4*e*) is apt to be more elongated than in the recent or Portuguese specimens (fig. 4*l*), and especially than in the Maryland (fig. 4*h*) specimens.

Terga very slightly beaked; the surface towards the carinal end of the valve, in some of the fossil specimens, is feebly striated longitudinally. There is either a slight depression (fig. 4*k*), or more commonly a deep longitudinal furrow (fig. 4*g*, 4*o*) with the edges folded in and touching each other, extending down the valve to the spur, and causing the latter to vary in width relatively to its length. When the furrow is closed in, the spur is about one fourth of the entire width of the valve, and has its lower end obliquely rounded, and stands at about its own width from the basi-scutal angle: when there is only a slight depression and no furrow (as is always the case with young specimens, and in the specimens (4*k*) from Maryland), the spur is broader, equalling one third of the width of the valve, with its lower end almost truncated, and standing at about half its own width from the basi-scutal angle. But the absolute length of the spur, also, varies considerably in the Coralline Crag specimens; it is often very long, (fig. 4*g*) compared to the whole valve. In many Italian specimens (4*o*) it is long and broad. The basal margin of the valve on the carinal side of the spur is sometimes slightly hollowed out; and when the longitudinal furrow is closed, this side slopes considerably towards the spur. Internally, the articular ridge and the crests for the tergal depressor muscles are moderately prominent.

Parietes, the longitudinal septa sometimes stand near each other, making the parietal pores small. The *radii* have oblique summits, but to a variable degree; their septa are unusually fine, and are denticulated on their lower sides; the interspaces are filled up solidly. The *alæ* have their summits very oblique, with their sutural edges nearly or quite smooth. In most of the fossil specimens (Tab. I, fig. 4*b*, *r*), and slightly in some of the recent specimens, the surface of the sheath presents an unusual character, in a narrow, longitudinal, slightly raised border, running along the sutures, on the rostral side of each suture.

Basis thin, porose; sometimes with an underlaying cancellated layer.

All the recent specimens which I have seen, were, with one exception, attached to various shells and crabs, and to each other. The tertiary specimens are often congregated together into great masses. Including the recent and fossil specimens, this species encircles the globe. During the miocene period it seems to have been the commonest existing Sessile Cirripede; now, it does not appear to be common, excepting, perhaps, at Panama.

5. Balanus porcatus. Tab. I, fig. 5*a*—5*g*.

Balanus porcatus. *Emanuel da Costa.* Hist. Nat. Test. Brit., p. 249, (1778).
Lepas balanus. *Linn.* Syst. Naturæ, (1767).
— — *Born.* Testacea Mus. Cæs. Desc., Tab. 1, fig. 4, (1780).
— — *Chemnitz.* Syst. Conch., 8 Band., Tab. 97, fig. 820, (1785).
Balanus arctica patelliformis. *Ellis.* Philosoph. Transact., vol. 50, Tab. 34, fig. 18, (1758).
— sulcatus. *Bruguière.* Encyclop. method., Tab. 164, fig. 1, (1789).
Lepas costata and Balanus. *Donovan.* British Shells, 1802–1804, Tab. 30, fig. 1, 2.
Lepas Scotica. *W. Wood.* General Conchology, Pl. 6, fig. 3, sed non *Lepas balanus*, Pl. 7, fig. 3, (1815).
Balanus angulosus. *Lamarck* (1818), in Chenu, Illust. Conch., Tab. 11, fig. 11.
— tesselatus. *Sowerby* (!) Mineral Conchology, Tab. 84, (1818).
— Scoticus. *Brown.* Illust. Conch. Great Britain, Pl. 7, fig. 2, sed non Pl. 6, fig. 9 et 10 (1827) : 2d edit., Pl. 53, fig. 1–3, 22, 23 et Pl. 54, fig. 1–3.
— geniculatus. *Conrad.* Journal Acad. Philadelphia, vol. vi, part 2, p. 265 (1830), Tab. 11, fig. 16.
— — *Aug. Gould* (!) Report on the Invertebrata of Massachussetts, fig. 9 (1841).

B. parietibus, sed non basi, poris perforatis; testâ albâ, plerumque longitudinaliter acutè costatâ; radiorum marginibus superioribus pæne basi parallelis: scuto longitudinaliter striato; tergi apice producto, purpureo.

Parietes, but not the basis, permeated by pores; shell white, generally sharply ribbed longitudinally; radii with their summits almost parallel to the basis. Scutum longitudinally striated; tergum with the apex produced and purple.

Fossil in the Glacial deposits of Scotland (Isle of Bute), of Uddevalla, and (Beaufort) Canada. In the Mammaliferous Crag (Bramerton, Thorpe) and Red Crag (Sutton); Mus. Lyell, J. de C. Sowerby, S. Wood, Bowerbank, &c.

Recent, England, Ireland, Scotland, Shetland Islands, Iceland, Davis's Straits, 66° 30′ N.; Lancaster Sound, 74° 48′ N. Maine and Massachussetts, United States. China (?) In deep water, commonly adherent on shells, crustacea, and rocks.

This species can be at once distinguished from all the foregoing by the basis being solid or not perforated by pores; and from all the following species, with the exception of *B. crenatus*, by the parietes having large square pores or tubes. From *B. crenatus*, this species can be distinguished by its longitudinally striated scuta, purple-beaked terga, and by the peculiar structure, immediately to be described, of its parietal pores; and in most cases even by its general aspect, larger size, and ribbed walls. When, however, *B. porcatus* and *crenatus* have grown together on the same irregular surface, for instance, on a Pecten, they sometimes resemble each other in a very deceptive manner. The opercular valves have not certainly been found fossil, but I have given drawings from recent specimens.

The parietes, (the basal margin of a small portion is represented at Tab. I, fig. 5*b*,) are perforated by large square longitudinal tubes: in the upper part these are filled up solidly without transverse septa; the longitudinal septa between the tubes are finely denticulated at their bases, and the denticuli extend unusually close to the outer lamina. In very young specimens the inner lamina of the parietes is ribbed, in lines corresponding with the longitudinal septa, as in the case of other species of the genus; but in medium and large-sized specimens, there are between such ribs from one to four smaller ribs, which do not correspond with any longitudinal septa; these are finely denticulated at their bases, and may be considered as the representatives of longitudinal septa which have not been developed and reached the outer lamina. I have seen no other instance of this structure, namely, the presence of a greater number of ribs on the inner (upper in fig. 5*b*) lamina of the walls than there are longitudinal septa. As the parietes rest on the basis, the circumference of the latter becomes marked in a very peculiar manner (fig. 5*c*), by the basal edges of the parietal septa. Hence the basis of this species can be distinguished from that of every other sessile cirripede: its circumference is plainly impressed by the main parietal septa which connect the inner and outer laminæ of the walls; and between these marks there are two or three smaller impressions of the so-called representative septa, which do not extend beyond the impression of the basal edge of the inner lamina. The upper surface of the middle part of the basis (more especially when slightly disintegrated) is faintly striated in radiating lines, of which the stronger lines are prolonged from the circumferential marks left by the main parietal septa, and the weaker lines from the marks left by the representative septa.

Dimensions.—The largest recent specimens which I have seen from great Britain or Ireland, have been 1·3 of an inch in basal diameter: in Mr. Cuming's collection, however, there is one much depressed specimen from the Shetland Islands, 2·1 in basal diameter; a regularly conical specimen from the coast of Massachussetts attains a nearly equal diameter. But out of the glacial deposits in the Isle of Bute, several specimens have this same diameter, namely, two inches, and are even more steeply conical, being 1·85 in height; some glacial specimens from Uddevalla and Canada, in Sir C. Lyell's collection, are 1·7 in basal diameter. Hence it appears, as we shall presently see is likewise the case with *B. crenatus* and *Hameri*, that northern specimens, and those from the United States and from the Glacial deposits, often exceed in dimensions those now living on the coasts of Great Britain and Ireland, or those found in the Crag.

This species is very common in the glacial deposits of Uddevalla, of Skien in Norway, and of Canada, and is associated with the same species, namely, *B. crenatus* and *Hameri*, as in the living state: I have seen, also, as just stated, specimens from the same formation in the Island of Bute, Scotland. I have examined numerous specimens from the Mammaliferous Crag, and a few from the Red Crag of England. I owe to the kindness of Mr. J. de C. Sowerby an inspection of the original specimens of *B. tesselatus* of the Mineral Conchology, which is certainly the present species.

6. Balanus crenatus. Tab. I, fig. 6*a*—6*g*.

B. crenatus. *Bruguière.* Encyclop. Method. (des Vers) 1789.
Lepas foliacea, *var. a.* *Spengler.* Skrifter af Naturhist. Selskabet, b. i, 1790.
— borealis. *Donovan.* British Shells, Pl. 163, (1802–1804).
B. rugosus. *Pulteney* (?) Catalogue of Shells of Dorsetshire, 1799.
— *Montagu* (?) Test. Brit., 1803.
— *Gould* (!) Report on Invertebrata of Massachussetts (1841), fig. 10.
B. glacialis (?) *J. E. Gray.* Suppl. Parry's Voyage, 1819.
B. elongatus (!) clavatus (!), *auctorum variorum.*

B. parietibus, sed non basi poris perforatis; testâ albâ; radiorum marginibus superioribus obliquis, asperis, rectis; scuto sine adductoris cristâ; tergi calcare rotundato.

Parietes but not basis permeated by pores; shell white; radii with their oblique summits rough and straight; scutum without an adductor ridge; tergum with the spur rounded.

Fossil in glacial deposits of Scandinavia and Canada, Mus. Lyell; in the mammaliferous and Red (Sutton) and Coralline Crags; Mus. S. Wood, J. de C. Sowerby, Bowerbank, &c. Miocene formation, Germany, Mus. Krantz.

Recent in Great Britain, Scandinavia, Arctic Regions as far as Lancaster Sound, in 74° 48′ N.; Behring's Straits; United States; Mediterranean; West Indies; Cape of Good Hope. Generally attached to shells and crustacea in deep water.

Under the last species I have shown that the porose parietes, but solid basis, distinguish this species easily from all the others, with the exception of *B. porcatus*, from which it can readily be known by the characters of its opercular valves, as already thereunder stated. Judging by external appearances alone, which ought never to be trusted to in the identification of any sessile cirripede, this species might easily be confounded with *Bal. dolosus*, found fossil in the same deposits.

This species presents a great diversity of external aspect: I have had figured (Tab. I, fig. 6*a*) one of the commonest appearances presented by it; but frequently the shell is quite smooth and depressed, or extremely much elongated and cylindrical, or even club-shaped. The *basis* is generally thin and slightly furrowed in lines radiating from the centre, but it is not permeated by pores; when, however, in large and old specimens it becomes thicker, as in Tab. I, fig. 6*c*, its edge is very distinctly pitted by little hollows, which might sometimes be easily mistaken for the orifices of pores: the absence of pores is a very important character in the diagnosis of *B. crenatus.* The basis is less firmly attached to the supporting surface than is usual with most cirripedes, and consequently it often separates from it together with the parietes. With regard to the opercular valves (6*d*—6*g*) drawn from recent specimens, I need here only state that the most conspicuous characters are the large articular ridge to the scutum, and the reflexed apices of all four valves, though this latter character is highly variable. I must refer to my Monograph on the Balanidæ for a full description of these valves.

The largest recent British specimen which I have seen was only ·55 of an inch in basal diameter: specimens from Greenland and the northern United States, frequently attain a diameter of three-quarters of an inch, and I have seen one single sowewhat distorted specimen actually 1·6 of an inch in basal diameter. Where individuals have grown crowded together, their length is often twice, and even occasionally thrice as great as their diameter; thus I have seen a recent Greenland specimen 1·6 of an inch in length, and only ·75 in diameter. This species, in its recent state, as may be seen under the habitats, has an enormous range. I have felt myself unwillingly compelled to admit that it ranges from the Arctic Regions in 74° 48′ N. to the Mediterranean, the West Indies, and Cape of Good Hope. That this species should live in the tropical seas is the more surprising, as the large size of the specimens in the northern seas and in the glacial deposits, might fairly have been supposed to have indicated special adaptation for a cold climate. This great geographical range, however, of the species accords with its range in time from the present day to the Coralline Crag period. The specimens from the glacial deposits which I have examined, chiefly in Sir C. Lyell's collection, are very fine and large, and appear, on an average, to attain as large or larger dimensions than the recent specimens from the United States; they are often associated, like the now living individuals, with *B. porcatus* and *Hameri*: they come from the well-known formation of Uddevalla, and from Canada. There are well-characterised specimens in the mammaliferous Crag, at Bramerton and near Norwich, in Sir C. Lyell's and Mr. Wood's collections, and from Sutton and other places in the Red Crag of the eastern shores of England: these specimens are not only smaller than the glacial, but than the recent English specimens; for the largest Crag specimens which I have seen had a basal diameter ·5 of an inch, ·3 to ·4 being their ordinary size. The specimens which I have seen from the Coralline Crag, and some others sent me by Krantz from the miocene formation of Flonheim bei Abzei, in Germany, had not their opercular valves, yet I cannot doubt, considering how few species there are having porose walls and a solid basis, that I have rightly identified these specimens as belonging to *B. crenatus*.

7. Balanus Hameri, Tab. I, fig. 7*a*—7*d*., Tab. II, fig. 1*a*, 1*b*.

Lepas Hameri. *Ascanius*. Icones rerum naturalium, Tab. 10, 1767.
— tulipa. *O. F. Müller*. Prodromus. Zoolog. Dan. 1776; sed non *L. tulipa*, in Poli, Test. ut Siciliæ; necnon *B. tulipa*, in Bruguière, Encyclop. method.; necnon *B. tulipa*, in Sowerby, Genera of Shells.
— tulipa alba. *Chemnitz*. Syst. Conch., Tab. 98, fig. 832.
— foliacea. *Spengler*. Skrivter af Naturhist. Selskabet, 1 B. 1790.

Balanus candidus. (Tab. emendata) *Brown.* Conch. Great Britain (1827), Tab. 6, figs. 9 and 10, and 2d edit. Tab. 54, figs. 9–12.
— tulipa. *Lyell.*[1] In Phil. Transact., 1835, p. 37, Tab. 2, figs. 34–39.

B. nec parietibus, nec basi, nec radiis poris perforatis; testâ albâ; radiorum marginibus superioribus obliquis, lævibus, arcuatis; aciebus suturalibus lævibus; scuto angusto longitudinaliter, debiliter striato; tergi calcare angusto, rotundato.

Parietes, and basis, and radii not permeated by pores; shell white; radii with their oblique summits smooth and arched; sutural edges smooth; scutum narrow, feebly striated longitudinally; tergum with the spur narrow, rounded.

Fossil in Red Crag (Sutton), Mus. S. Wood Doubtfully in the Glacial beds of Scotland. In the Glacial deposits at Uddevalla, in Sweden; and Beaufort, Canada, Mus. Lyell. Banks of the Dwina, Russia, Mus. Murchison. Greenland, "in blue clay," according to Spengler.

Recent on the Coast of Yorkshire; Scotland; Galway, Ireland; Isle of Man, and Anglesey, twelve fathoms. Generally in deep water; not very common. George's Bank, Massachussetts, United States. Iceland, Finmark, and the Faroe Island, according to Spengler. Attached to crustacea, mollusca, stems of fuci, and stones; often associated with *B. porcatus* and *crenatus*.

I have seen, in Mr. Wood's collection, from the Red Crag, fine and perfectly preserved specimens of a rostrum, and of a lateral compartment. The latter was three inches in height, and, including the alæ, one inch in width. I have also seen a specimen said doubtfully to have come from the glacial beds of Scotland. As it is so very common in the deposits of this same age in other countries, no doubt it will hereafter be found more plentifully in Scotland, and probably in the mammaliferous Crag of England. *Balanus Hameri* is a very fine species; I have seen a recent specimen from the coast of Yorkshire, two inches in diameter, and one inch and three-quarters in height: another specimen was three inches in height. The specimens in the glacial deposits, seem to have acquired larger dimensions: a compartment from Uddevalla being nearly four inches in height. The white colour, smooth surface, and regularly arched radii, give to the shell a very elegant appearance, which has appropriately been compared to that of a white tulip. The diagnosis of this species is easy; the walls as well as the basis being solid or not porose, serve to distinguish it from all other forms except certain varieties of *Bal. unguiformis,* and by several minor characters, such as the finely striated and more elongated scuta, &c., *Bal. Hameri,* can be discriminated from *B. unguiformis.*

[1] Sir C. Lyell remarks that this is apparently the *B. Uddevallensis,* (Linn.), of Swedish lists of fossils. Prof. E. Forbes has shown ('Mem. Geolog. Survey of England,' vol. i, p. 364) how this name arose, from a short description, prior to the introduction of the binomial system, "Lepas quæ Balanus Uddevallensis," given by Linnæus in his Wast-Gotha Resa, in 1747.

For the reference to Ascanius' work, which is on the binomial system, and subsequent to the 10th edit. of Linnæus in 1758, I am greatly indebted to Mr. Sylvanus Hanley. Had it not been for this gentleman, I should have used Müller's name of *B. tulipa* as the first name.

In the recent condition, the compartments of the dead shell fall apart with singular facility; and Sir Charles Lyell has remarked ('Philosophical Transactions,' 1835, p. 37) that in the glacial deposits of Scandinavia, the shell is never found whole, but the separated compartments in abundance: it appears, also, that the basis likewise easily separates from its support. The extreme edge of the basis is finely crenated, and not pitted as in *Bal. crenatus;* the crenations or teeth are produced by the edge of the basis fitting in between the longitudinal septa on the internal surface of the parietes. There is one peculiarity in the alæ of this species in its recent state, which I have observed in no other species, and which can be distinguished in some of the fossil specimens, as in Tab. II, fig. 1*b*,—namely, the presence of an excessively fine linear furrow running along the sutural edge, a little towards the inner side, and filled (in the recent state) with a yellow ligamentous substance.

In regard to the opercular valves, (Tab. I, fig. 7*a*—7*d*, drawn from recent and glacial specimens,) I need here only mention, that in the *scuta,* their flatness, elongation, and delicate longitudinal striæ, are their chief characteristics. In very old and large specimens of the terga (as in the specimen, fig. 7*d*, figured from Uddevalla), the basal margin on the carinal side of the spur slopes down towards it in a remarkable manner.

8. Balanus bisulcatus, Tab. II, fig. 2*a*—2*h*.

Balanus sulcatinus (?) *Nyst*, apud D'Omalius (sine descript. aut tabulâ), Géologie de Belgique, 1853.[1]

B. nec parietibus, nec radiis poris perforatis; basi poris magnis perforatâ; radiorum marginibus superioribus obliquis, lævibus; aciebus suturalibus lævibus; scuto angusto, sulcis longitudinalibus 2 ad 4; tergi calcare brevissimo dimidiâ valvæ latitudine.

Neither walls nor radii permeated by pores; basis permeated by large pores; radii with their upper margins oblique and smooth; sutural edges smooth; scutum narrow, with from two to four longitudinal furrows; tergum with the spur very short, broad as half the valve.

Var. plicatus (fig. 2*c*), *with the walls deeply folded; radii narrow, with their upper margins very oblique.*

Fossil in Coralline Crag; Ramsholt, Gedgrave, Sutton; Mus. S. Wood, Bowerbank, J. de C. Sowerby. Rauville, dans le Cotantin, Mus. G. B. Sowerby. *Var. plicatus,* Coralline Crag, Sutton, Mus. S. Wood, Bowerbank. Bolderberg, near Hasselt, Belgium, Mus. Bosquet.

[1] I am indebted to M. Bosquet for a specimen, bearing this name and reference, found in the 'Système Bolderien' of Dumont, (miocene according to Sir C. Lyell) at Bolderberg. The specimen consists of a rostrum, with a portion of the base attached; and as these parts are in some degree characteristic, I fully believe this specimen to be *B. bisulcatus.*

General Appearance.—Shell (fig. 2*a*) conical or tubulo-conical, often rather globose; walls frequently thin, either very smooth, or deeply plicated longitudinally: occasionally the same specimen is smooth in the upper part (fig. 2*b*), and strongly plicated in the lower. The radii in the large specimens are wide, and with their upper margins only slightly oblique; in the smaller they are narrower, and much more oblique; but in each case their upper margins are smooth and slightly bowed. Colour apparently originally nearly white, but with the alæ generally, in the smaller specimens, clouded with a dark tint: the radii are usually striped feebly in longitudinal lines. Basal diameter of largest specimen ·8 of an inch; but this seems to have been an unusual size.

Scuta: (fig. 2*e*) narrow, with the basal margin forming an unusually small angle with the occludent margin; surface slightly convex, with lines of growth approximate, moderately prominent; on the tergal half of the valve, two distinct rather broad furrows, with sometimes a third, and even a fourth, nearer to the occludent margin, extend from the apex down the valve, and give it a very peculiar appearance: the furrows near the tergal margin are the deepest. Internally (fig. 2*g*), the upper part of the valve is roughened with small points: the articular furrow is unusually wide: the articular ridge is very prominent and but little reflexed, with the lower end almost abruptly cut off: the adductor ridge is prominent, but short: there are small deepish pits for the rostral and lateral depressores.

Terga (fig. 2*f*), broad, flat, with a slight narrow prominent rim along the scutal margin, which margin is slightly bowed. The basal margin on the carinal side of the spur slopes so gradually towards the spur, that the latter is barely distinct, and is very short, not depending nearly half its own width beneath the basi-scutal angle: the spur, also, is broad, namely, measured across the upper part, as broad as half the valve; its basal end is obliquely rounded off on the carinal side; it is placed close to basi-scutal angle. The carinal margin of the valve is just perceptibly bowed, and is formed by rectangularly upturned lines of growth. Internally (fig. 2*h*), the upper part of the valve is rough; the articular ridge is prominent; the crests for the tergal depressores muscles are moderately well-developed.

Parietes, not porose; internally, the ribs are smooth, with their basal edges very finely or barely denticulated. The *radii* (as already stated) are of variable breadth; they have their upper margins either very slightly or highly oblique, but always smooth and rounded: their sutural edges are quite smooth, or sometimes, with a strong lens, traces of transverse striæ, representing septa, can just be discovered. The *alæ* have their upper margins very oblique; their sutural edges are, in the large specimens, quite smooth; in the younger ones, plainly crenated; the recipient furrow being clearly marked by the teeth. *Basis* plainly porose.

Varieties.—It is certain (fig. 2*b*) that there are longitudinally plicated specimens of this species, and that the obliquity of the upper margins of the radii also varies a little; nevertheless some of the deeply plicated specimens (fig. 2*c*) undoubtedly have a very different aspect from the ordinary varieties, and do really differ in the sutural edges of the alæ being crenated, and in the greater narrowness and obliquity of the radii; but these points are all commonly variable. I have not seen any large specimens of the variety (fig. 2*c*), *plicatus*, so as to compare them with the large specimens of the normal form, yet I can hardly entertain any doubt, considering their agreement in so many important points, that I have rightly treated these forms as mere varieties; it is unfortunate that none of the specimens of the *var. plicatus* seen by me have had opercular valves, as their presence would have removed all shadow of doubt. I have given a drawing, enlarged seven times, of some very young shells (fig. 2*d*), adhering in numbers on *Pecten Gerardii*, which I believe belong to the plicated variety of our present species, but which are much too young to be identified with certainty.

Affinities: this is a strongly characterised species, and nearly allied only to the following species, *B. dolosus*. The furrows on the scuta in some degree resemble those on the recent *B. lævis*, but there is no alliance with that species. It is certain that amongst recent species, the chief affinity is with *B. Hameri* and *amaryllis*.

9. Balanus dolosus. Tab. II, fig. 3*a*—3*d*.

B. nec parietibus, nec radiis poris perforatis; basi poris magnis perforatâ, radiorum marginibus superioribus obliquis, lævibus; aciebus suturalibus item lævibus; tergi calcare non admodum brevi, ⅓ valvæ latitudine.

Neither walls nor radii permeated by pores; basis permeated by large pores; radii with their upper margins oblique and smooth; sutural edges smooth; tergum with the spur not very short, broad as one third of valve.

Fossil in Red (Sutton) and Mammaliferous Crag; Mus. S. Wood, Bowerbank, Lyell, J. de C. Sowerby, Henslow, &c. Mammaliferous Crag, Postwick, near Norwich, Mus. Lyell.

This species so closely resembles *B. bisulcatus*, both externally and in all the essential characters of the parietes, radii, and basis, that it is quite superfluous to describe again these parts. The specific characters are derived from the opercular valves, which present well defined distinctions, found by me constant in several specimens of both species. *B. dolosus*, like *B. bisulcatus*, has quite smooth and deeply plicated varieties, often adhering to the same univalve. The ribs on the inner surfaces of the parietes are remarkably prominent, as shown in the drawing (fig. 3*a*) of the inside of the rostrum. I think the upper margins of the radii are in this species rather more oblique than in *B. bisulcatus*. The sutural edges of the radii are marked by the finest striæ, representing septa. The sutural edges of the alæ are generally distinctly crenated. The basis is often slightly cup-formed, and very plainly porose (fig. 3*b*): its upper surface is marked by radiating ridges: the septa between the radiating pores are themselves often in part porose, as was plainly the case in the specimen (fig. 3*b*) engraved. The orifice of the shell is large and elongated in its rostro-carinal axis, especially in young specimens. The basal diameter of the largest specimen is ·4 of an inch.

The *scuta* (fig. 3*c*) have no trace of the two or three longitudinal furrows so conspicuous on these valves in *B. bisulcatus*, and which, in that species, run down to the basal margin from the apex of the valve, this fact showing that the furrows occur in quite young individuals. The whole valve is not quite so narrow as in *B. bisulcatus*, but otherwise agrees with it in shape: internally, there is hardly any difference; but the articular furrow is not quite so wide: the articular ridge is very prominent, and abruptly truncated at its lower end: the adductor ridge is also prominent; it here runs a little higher up the valve than in *B. bisulcatus*. The *tergum* (fig. 3*d*) differs more in the two species: the spur is not so broad; measured in its upper part, it is only about one third of the entire width of the valve, instead of being half as wide as the valve: it is considerably longer, depending beneath the basi-scutal angle more than half its own width: the basal margin of the valve on the carinal side, does not slope so gradually into the spur; the occludent and carinal margins are slightly arched, as in *B. bisulcatus*. Internally, the surface is rough, the articular ridge is prominent, and the crests for the tergal depressores are well developed,—all as in *B. bisulcatus*. It is remarkable how generally the opercular valves have been preserved in this species in its fossil condition, as compared with most other species of the genus.

It is not easy to distinguish, by external characters, the rugged varieties of this species from *B. crenatus;* indeed, the only difference is that the furrows receiving the edges of the radii, generally, exhibit in *B. crenatus* slight impressions of the septa, which are entirely absent in *B. dolosus*. By internal characters, such as the non-porose parietes, and porose basis, our present species widely differs from *B. crenatus*.

10. Balanus unguiformis. Tab. II, fig. 4*a*—4*f*.

Balanus unguiformis. *J. de C. Sowerby* (!) Mineral Conchology (sine descriptione), Tab. 648, fig. 1, (Jan. 1846.)

— erisma. *J. de C. Sowerby* (!) Ib., fig. 2.

— perplexus. *Nyst*, apud D'Omalius (sine descript. vel Tab.), Géologie de la Belgique, 1853.[1]

B. parietibus tenuibus, interdum poris perforatis: radiis sine poris, marginibus superioribus obliquis; aciebus suturalibus tenuissimè crenatis: basi sine poris: tergi calcare angusto, obtuso.

[1] I am much indebted to M. Bosquet for specimens bearing this title, from Klein Spauwen, which certainly appear to me, as far as can be judged by the separated compartments, without the opercular valves. to belong to our present species.

Parietes thin, sometimes permeated by pores; radii without pores, with their upper margins oblique; sutural edges very finely crenated; basis without pores. Tergum with the spur narrow, bluntly pointed.

Var. erisma (fig. 4*b*), *with the walls longitudinally folded or ribbed.*

Fossil in the Eocene formations, Isle of Wight, Colwell Bay; Hordwell; Barton, (Chama Bed); Headon; Bembridge; Bergh, near Klein Spauwen, Belgium (?). Attached to various shells and wood. Mus J. de C. Sowerby, E. Forbes, F. Edwards, Charlsworth, T. Wright, Bowerbank, Tennant, Bosquet.

This species, the most ancient one as yet well known in the genus, presents to the systematist a most unfortunate peculiarity, in the parietes being almost as often as not permeated by small pores: I have seen no other instance, except to a limited degree in the recent *B. glandula*, of this character being variable, and hence it must be still considered of high classificatory value, in so varying a genus as Balanus. Owing to the kindness of Mr. F. Edwards, I have seen the original specimens, excellently figured by Mr. J. de C. Sowerby, in the 'Mineral Conchology,' under the names of *B. unguiformis* and *erisma*, between which I can perceive no difference, excepting that the walls in the latter are longitudinally folded,—a character we know to be variable in many species. In both varieties, the parietes are sometimes porose and sometimes solid. The smaller specimens, however, figured in the 'Mineral Conchology' to the right hand of the Plate, may possibly be a distinct species, as I infer from the narrowness of their radii. This species is intimately allied to *B. varians*, a fossil from the ancient tertiary plains of Patagonia. It is also allied to the recent *B. crenatus* and *glandula*.

General appearance.—Shell (fig. 4*a*), tubulo-conical, sometimes even sub-cylindrical: surface either very smooth, or slightly folded, or deeply folded so as be strongly ribbed longitudinally: orifice rather large, rhomboidal, narrow at the carinal end, toothed, but not deeply: walls rather thin and fragile: radii of moderate width, with their summits oblique, not quite smooth. Basal diameter of largest specimen about three quarters of an inch.

Scuta (fig. 4*c*), with the external surface smooth: there is a trace of a furrow running down the valve from the apex, near to the occludent margin, and this is only worth mentioning from the analogous furrows in *B. bisculatus*. Internally (fig. 4*e*), the upper surface of the valve is roughened: the articular ridge is very prominent, and slightly reflexed: there is no distinct adductor ridge; there is a slight but variable depression for the lateral depressor. *Tergum* (fig. 4*d*), with the longitudinal furrow shallow; spur moderately long, about one fourth or one fifth of the width of the valve; placed at about its own width from the basi-scutal angle; basal end bluntly pointed; the basal margin on the opposite sides of the spur forms a nearly straight line; the carinal margin has an extremely narrow border formed by upturned lines of growth. Internally (fig. 4*f*), the surface is roughened with little points: the articular ridge is prominent: the crests for the tergal depressores moderately prominent.

Parietes: the longitudinal ribs on the internal surface are either feebly, or, in the lower part, strongly developed; their basal ends are only just perceptibly denticulated. As already stated, in about half the specimens, there were no traces of parietal pores; in the other half there were either distinct or obscure pores; the pores are circular, generally of unequal sizes, and never large; in the same individual they would sometimes be almost wholly absent in some of the compartments, and quite plain in the other compartments. The *radii* are either moderately wide or rather narrow, and have their upper margins very oblique, and not distinctly arched, and not quite smooth: their sutural edges are very finely crenated, the teeth or septa not being denticulated. The upper margins of the *alæ* are rather less oblique than those of the radii: their sutural edges are barely crenated. The *basis* is thin, and without any trace of pores; the upper surface is sometimes furrowed in radiating lines.

11. Balanus inclusus. Tab. II, fig. 5*a*—5*g*.

B. nec parietibus, nec radiis poris perforatis; basi poris perforatâ: testâ rufo-fuscâ: radiis latis, marginibus superioribus aut non obliquis aut modicè; aciebus suturalibus cum septis planè denticulatis: scuto sine adductoris cristâ: tergi calcare subangusto.

Neither walls nor radii permeated by pores; basis porose; shell reddish-brown; radii broad, with their upper margins not oblique, or only moderately oblique; sutural edges with plainly denticulated septa: scutum without an adductor ridge; tergum with the spur rather narrow.

Var. (*a*) (fig. 5*c*, 5*d*), *with the shell elongated in its rostro-carinal axis; basis narrow, clasping the stem of a zoophyte; lateral compartments much broader than the almost linear rostrum, carina, and carino-lateral compartments.*

Var. (*b*), *with rough longitudinally folded walls, and with the summits of the radii forming an angle of about* 45° *with the basis.*

Fossil in Coralline Crag; Sutton and Gedgrave; attached to foliaceous Bryozoa; Mus. S. Wood, Bowerbank. *Var. a*, Coralline Crag, Sutton, attached to cylindrical branches of corals; Mus. S. Wood, Bowerbank. *Var. b*, attached to shells, Osnabruck, Hanover, Mus. Lyell; Bunde, Westphalia, Mus. Krantz.

My materials consist of a beautiful series of specimens in Messrs. Wood and Bowerbank's collections; but unfortunately only a single young specimen had its opercular valves preserved. Not one specimen of the very curious variety (*a*) had opercular valves, yet I cannot feel any doubt about its being only a variety, caused by its attachment to a thin cylindrical branch of a coral, instead of to a foliaceous Bryozoon; it will, however, be convenient to give a separate description of this very remarkable form. With respect to var. (*b*), both sets of specimens came to me from the Continent, with the name of

B. stellaris, of Bronn; but as Bronn distinctly states, that in his species the parietes are porose, and as such is not here the case, this cannot possibly be that species: these specimens did not possess their opercular valves, and therefore cannot be identified with certainty.

General Appearance.—Shell conical (fig. 5*a*, 5*b*), with the orifice rather large, and rhomboidal. The surface is very smooth, except in var. (*b*) from the Continent, in which it is rugged and longitudinally folded. The colour is ochreous-brown (chiefly no doubt derived from the imbedding substance), tinged with red. The radii often have a much darker and more distinct red tint; they are sometimes longitudinally striped with dirty white. The radii are broad, with their summits straight, and very slightly oblique; in *var. b*, however, they slope at an angle of about 45°. Basal diameter of largest specimens ·6 of an inch; but this is an unusual size.

Scuta, with the growth ridges little prominent. Internally (fig. 5*f*, from a young individual) the articular ridge is moderately prominent, with its lower end very obliquely rounded off; there is no adductor ridge; there is a minute pit for the lateral depressor muscle. *Terga*, with a slight longitudinal depression extending down to the spur; spur short, with its lower end almost square or truncated, about one fourth of width of valve, and placed at about half its own width from the basi-scutal angle. Internally (fig. 5*g*), the articular ridge is prominent; the crests for the tergal depressores are feebly developed.

Parietes, moderately thick and generally strongly ribbed internally, without parietal pores. *Radii*, wide, with their upper margins straight, not smooth or rounded, and very slightly (or, in *var. b*, moderately) oblique; their sutural edges have well-developed septa, which are denticulated: the interspaces between the septa are filled up solidly. The *alæ* have their upper margins oblique: they are only slightly, and sometimes not at all, added to above the level of the opercular membrane: their sutural edges are smooth. The *basis* is thin, but plainly porose.

Var. (*a*) (fig. 5*c*, 5*d*).—With respect to this remarkable variety, any one would at first think it specifically distinct. The shell is much compressed, or elongated in the rostro-carinal axis, sometimes to a great degree; I have seen a specimen ·25 of an inch in this axis, and only ·1 in its broadest part; but this is a very unusual degree of elongation. The most remarkable character is the extraordinary narrowness of the carina, of the carino-lateral compartments, and of the rostrum, compared with the great breadth, especially along the basal margin (fig. 5*d*), of the lateral compartments. The radii are of unusual breadth. The tips of the rostrum and of the lateral compartments are a little arched in, tending to make the shell somewhat globular. The true basis is extremely narrow (fig. 5*d*): it is deeply grooved, from clasping the thin, cylindrical stem of the coral to which it has adhered; and I have seen specimens in which the opposite edges of the groove had met, a tube having been thus actually formed. From the grooved basis, and from the elongation of the shell in the rostro-carinal axis, this variety presents so close a general resemblance to *Balanus calceolus*, and its allies, that I have seen it in a collection arranged on the same tablet with a fossil specimen of *B. calceolus.* Notwithstanding the above several strongly-marked characters, by which this variety differs from the ordinary form, there is a resemblance in colour and aspect, which, though difficult to be described, made me from the first suspect that the two were specifically identical. In no point of real structure is there any difference, excepting that, perhaps, the pores in the basis are here rather smaller; but this might arise from the little development of the peculiar basis. Having come to this conclusion, I was interested by finding a specimen (fig. 5*e*) in Mr. Wood's collection, which had originally fixed itself (judging from the form of the basis) on a thick cylindrical stem, but which had subsequently grown on to an adjoining flat surface; consequently, one side of the shell presented all the peculiar characters of the present variety, but not strongly pronounced, whereas the other side, at the rostral end, was undistinguishable from the ordinary form. The unequal development of the rostrum on the two sides was very striking, and clearly showed how great an effect could be produced by the nature of the surface of attachment.

This singular variety cannot be considered accidental, in the sense in which this term may be applied in some cases: the pupa evidently fixes itself intentionally, in a certain definite position, on the branch of the coral (when a branch is chosen), exactly as in the case of *Balanus calceolus*, or *Scapellum vulgare*,—species which always live attached to branches. But when other Balani occasionally fix themselves on branched corals, their position seems to be accidental and unsymmetrical; thus among the symmetrically elongated specimens of the present species, I found one specimen of *Balanus bisulcatus*, which had evidently been attached in an almost transverse position to a branch, and had thus become much distorted; so, again, I have seen specimens of the recent *B. amaryllis* attached irregularly to a Gorgonia, in the midst of the symmetrically elongated shells of *Balanus navicula*, an ally of *B. calceolus*.

This variety does not seem to attain so large a size as the ordinary form.

Affinities.—This species is allied to *B. unguiformis* and *B. varians*, but is perhaps more nearly related to the recent *B. allium*, an inhabitant of the Barrier Reef of Australia. The longitudinally folded variety (*b*) can hardly be distinguished by external aspect, or even by the opercular valves, from *B. crenatus;* but when the shell is disarticulated, the porose walls and non-porose basis of *B. crenatus*, allow of no mistake in the diagnosis of the two species.

Sub-Genus—Acasta.

Acasta. *Leach.* Journal de Physique, tom. lxxxv, 1817.

Valvæ testæ 6; *parietes et basis non porosa; basis calcarea, cyathiformis, non elongata. Valvæ operculares inter se articulatæ, subtriangulares. Spongiis, aut rarò Isidis cortici, affixa.*

Compartments six; parietes and basis non-porose: basis calcareous, cup-formed, not elongated, attached to Sponges, or rarely to the bark of Isis: scutum and tergum articulated together, subtriangular.

Under the last genus, I have made a few remarks on the close affinities of this sub-genus to Balanus, and have given my reasons for retaining it, so that I need not here repeat them.

ACASTA UNDULATA. Tab. II, fig. 6*a*—6*f*.

A. testâ, ad speciem, ut in "A. spongites," sed majore: scuto externe striis longitudinalibus, sæpe binis, signato, sulcis intermediis latioribus: tergi calcare, pæne ½ valvæ latitudine.

Shell, apparently, as in A. spongites, but larger: scutum marked by longitudinal ridges, often in pairs, with the intermediate furrows rather wide: spur of tergum nearly half as wide as valve.

Fossil in Coralline Crag (Sutton), Mus. S. Wood, Bowerbank.

I owe to Mr. Wood the inspection of a fine suite of valves, which, though separate, I have no reason to doubt have all been rightly attributed to the same species. Owing to the shell never having been found entire, its general shape is not known, and, what is of more consequence, the relative proportional width of the parietes of the carino-lateral compartment is unknown. I have (but with doubt) given it a distinct specific name, owing to the peculiar character of the furrows on the scuta, and to the large size of the whole shell. In its other characters it comes nearest to *A. spongites*, excepting in the spur of the tergum, which resembles that of *A. sulcata*.

The external surfaces of the compartments appear generally to have been smooth; but in several specimens they are studded with the sharp shelly points so common in the genus. A rostrum (Tab. II, fig. 6*a*), and lateral compartment (fig. 6*b*), have been figured. The radii are not wide. The basis (fig. 6*c*) is cup-formed: its edge is either quite smooth, or is very finely crenated. The basis is sometimes quite irregularly perforated, as in the case of several recent species, by numerous minute orifices, which, when the animal was alive, no doubt were covered by membrane. Internally the parietes are feebly ribbed, as in *A. spongites*. Judging from the dimensions of the separated valves, this species must have equalled and perhaps exceeded in size the largest living species, namely, *A. glans*, from Australia. Hence we may infer, that the basal diameter probably exceeded ·55 of an inch: I may add, that the largest European specimens of *A. spongites*, from Naples and Portugal, are only ·3 of an inch in basal diameter.

Scuta (fig. 6*e*).—These seem to resemble the scuta of *A. spongites* in all respects, except in the external longitudinal ridges standing much further apart, and, consequently, in the furrows being much wider: each ridge is generally double. Although there is a good deal of variability in the character of these ridges in *A. undulata*, and likewise in *A. spongites*, I have not seen any form intermediate between them. It must, however, be confessed, that this is an extremely variable character in many sessile cirripedes. Internally the scutum (fig. 6*d*) is chiefly characterised by the absence of characters, that is, by the slightness of the pits for the muscles, and the little prominence of the articular ridge. In the *tergum* (fig. 6*f*), the spur is about half the width of the whole valve, and therefore rather wider than in *A. spongites.*

Genus—PYRGOMA.

PYRGOMA. *Leach.* Journal de Physique, tom. 85, 1817.
BOSCIA. *Ferussac.* Dict. Classique d'Hist. Naturelle, 1822.
SAVIGNIUM. *Leach.* Zoological Journal, vol. ii, July, 1825.
MEGATREMA. *Ib.* Ib.
ADNA. *Ib.* Ib.
DARACIA. *J. E. Gray.* Annals of Phil. (new series), August, 1825.
CREUSIA. *De Blainville.* Dict. Sc. Nat., Pl. 116, 1816-30.
NOBIA. *G. B. Sowerby, junr.* Conchological Manual,[1] 1839.

Valvæ testæ in unam confluente: basis cyathiformis aut subcylindrica, coraliis affixa: valvæ operculares inter se articulatæ.

Shell formed of a single piece: basis cup-formed, or subcylindrical, attached to corals: scutum and tergum articulated together.

This genus can at once be recognised by the shell consisting of a single piece without sutures, whether viewed externally or internally, and by the cup-shaped basis, attached and often imbedded in corals. The one species, *P. Anglicum*, found both recent and fossil, together with a closely allied recent species, *P. Stokesii*, in all the characters derived from the opercular valves, closely resemble Balanus and other ordinary forms, and for this very reason they have some slight claims to be generically separated from the other species of Pyrgoma; for in these latter, the opercular valves seem to have broken loose from all law, presenting a greater diversity in character than do all the other species of Balaninæ and and Chthamalinæ taken together.

[1] The name, Nobia, is given in this work on the authority of Leach, but this must be a mistake, probably caused by some MS. name, (that fertile source of error in nomenclature), having been used.

Pyrgoma Anglicum. Tab. II, fig. 7*a*—7*c*.

Pyrgoma Anglica. *G. B. Sowerby.* Genera of Recent and Fossil Shells, fig. 7, No. 18, Sept. 1823 (sine descript.).
Megatrema (Adna) Anglica. *J. E. Gray.* Annals of Philosoph. (new series), vol. x, Aug. 1825.
Pyrgoma sulcatum. *Philippi.* Enumeratio Molluscorum Siciliæ, Tab. 12, fig. 24, (1836).
— anglica. *Brown.* Illustrations of Conchology, (2d edit., 1844), Tab. 53, fig. 27—29.

P. testâ abruptè conicâ, purpureo-rubrâ; orificio ovato, angusto; basi porosâ, plerumque è coralio exserta: scuto et tergo subtriangularibus.

Shell steeply conical, purplish red: orifice oval, narrow: basis permeated by pores, generally exserted out of the coral: scutum and tergum sub-triangular.

Fossil in the Coralline Crag (Ramsholt) Mus. S. Wood.
Recent on the south coast of England and of Ireland, (12 to 45 fathoms, Forbes and MacAndrew); Sicily; Madeira; St. Jago, Cape de Verde Islands; generally attached to the edge of the cup of a Caryophyllia, in deep water, but at St. Jago within the tidal limits.

I have considered this fossil as identical with the recent species, but, as may be seen from the following description, it presents several slight differences; yet they are such that I dare not found a new species on only a few specimens thus characterised.

The shell is steeply conical, slightly compressed, with the lower part having rounded, approximate, radiating ribs; these ribs seem to be more prominent in the fossil than in the recent specimens. Colour dull purplish-red. Orifice oval, small, and narrow. The basis is not deeply conical, and occasionally is even flat: in the Crag specimens it is almost wholly imbedded in the coral to which it is attached; but in recent specimens it is generally exserted. Externally the basis is furnished with ribs corresponding with those on the shell. The largest recent specimens which I have seen, from St. Jago, was ·22 of an inch in basal diameter; but some few of the British specimens are nearly as large, and one of the fossils from the Coralline Crag a very little larger.

The *scuta* and *terga* are of the ordinary shape of these valves in Balanus and its allies. *Scutum* (fig. 7*b*) triangular, with the basal margin a little curved and protuberant; adductor and articular ridges distinct from each other, moderately prominent; there is a small hollow for the lateral depressor muscle: in the fossils, the adductor ridge (as figured) is more distinct from the articular ridge, and consequently the cavity for the lateral depressor muscle is wider and less deep that in recent specimens. *Tergum*, I have not seen a fossil specimen, but have figured a recent valve (7*c*); it is triangular, with the spur rather narrow, moderately long, placed near, but not confluent with, the basi-scutal angle of the valve. The basal margin forms an angle rather above a right angle with the spur. Internally the articular ridge and crests for the depressor muscles, feebly developed.

Internal Structure of the Shell and Basis.—Internally, the shell is ribbed, more or less prominently. The lower edge of the sheath, which is reddish, and extends far down the walls, seems always to project freely. In several specimens there were on each side, at the carinal end of the shell, a trace apparently of a suture, which could be perceived only on the sheath. The basis appears always to be permeated by minute tubes or pores, though these are sometimes rather difficult to be seen.

Michelotti, in the 'Bulletin Soc. Géolog.' Tom. 10, p. 141, has named, but not described, a species, viz., *Pyrgoma undata*, from the northern Italian Tertiary Strata.

Genus—Coronula.

Coronula. *Lamarck.* Annales du Muséum, tom. i, (1802).
Diadema. *Schumacher.* Essai d'un Nouveau Syst., &c., 1817.
Cetopirus (sed non Coronula). *Ranzani.* Memoire di Storia Naturale, (1820).
Polylepas. *J. E. Gray, (Klein).* Annals of Philosophy, (new series), vol. 10, 1825.

Valvæ testæ 6, æquali latitudine; parietes tenues, profundè plicati, plicis cavitates infrà solùm apertas efficientibus; valvæ operculares non inter se articulatæ, orificio testæ multo minores: basis membranacea. Cetaceis affixa.

Compartments six, of equal sizes: walls thin, deeply folded, with the folds forming cavities, open only on the under side of the shell: opercular valves much smaller than the orifice of the shell; when both present not articulated together: basis membranous. Attached to Cetaceans.

The structure of the shell of Coronula is complicated, and has been generally quite misunderstood. Without a long description and several figures it would be impossible to give a true idea of its singular structure; but, in order to make the following description at all intelligible, I must make a few remarks. The wall of each compartment, and therefore of the whole shell, is extremely thin; but strength is gained by its being folded in a very complicated manner, as may be seen in the rostral compartment, Tab. II, fig. 8*b*, by tracing the wall *e* to *e*′, to *e*″; the folds at their outer ends are elongated into transverse loops, the extremities of which touch each other; consequently, what appears to be the outside of the shell consists only of a portion of the wall, namely, the outsides of the transverse circumferential loops, together with the radii. These loops appear externally like much flattened longitudinal broad ribs. On the other hand, the inside of the shell, in which the body is lodged, consists of the inner ends of the folded walls, lined by the sheath, and by the alæ. The basal edges of the folded walls, in the line of the ray of the circular shell, are oblique; the outer ends, or transverse circumferential loops, having grown downwards at a greater rate than the inner ends. Between each fold of the walls, there is a flattened cavity, open at the bottom of the shell, and running up to the apex: these cavities are quite external to the cirripede, and are occupied by the epidermis of the whale to which the Coronula is attached: homologically they are only deep longitudinal furrows, and they would still have been furrows, had not the transversely elongated ends of the folds, *i. e.*, the circumferential loops, in all cases, after early growth, grown into close contact. The ends of these loops are generally locked together by rows of minute teeth. In all the species, when young, the wall of each compartment is folded three times, and therefore the whole shell has eighteen folds.

The radii, normally, are only part of the wall, modified by growing against an opposed compartment; and hence the radius in Coronula would have been extremely thin, like the wall, and the sutures between the six compartments excessively weak, had not the radii been specially thickened by numerous sinuous denticulated plates, springing from the inner lamina of the true radius, and running downwards, attached to the folded wall of the compartment to which the radius belongs, and with their free edges pressed against the folded wall of the opposed compartment. Hence the radii may be said to be compound. For the sake of strengthening the sutures, the alæ, also, are very unusually thick: but, notwithstanding their thickness and the thickness of the compound radii, owing to the depth of the folds of wall, they are separated from each other by a considerable space, and the alæ, instead of resting in chief part, as they should do, on the inner lamina of the radius, have to rest on special plates, developed apparently from the sheath. In the upper part of the shell, between the special plates on which the alæ rest, and the compound radii, there are in two of the three recent species, open chambers, six in number, occupied by the ovarian cæca; but in the fossil *C. barbara* these chambers are almost filled up solidly by shell. I hope that the terms used in the following description may be now in some partial degree rendered intelligible.

Coronula barbara. Tab. II, fig. 8*a*—8*e*.

Coronulites diadema (?) *Parkinson*. Organic Remains (1811), vol. iii, p. 240, pl. 16, fig. 19.

C. testâ (probabiliter) coroniformi, costis longitudinalibus convexis, aciebus earum crenatis, superficie internâ et externâ cristis transversis asperâ; radiis modicè crassis; spatio inter radios et alas solidè impleto.

Shell (probably) crown-shaped, with longitudinal convex ribs, having their edges crenated, and their surfaces rugged, both externally and internally, with transverse ridges: radii moderately thick; the space between the radii and the alæ solidly filled up.

Fossil in Red Crag, (Bawdsey and Sutton); Mus. S. Wood and Geological Society.

This species, though closely allied to *C. diadema* and easily confounded with it, I have no doubt is distinct. I owe to the kindness of the Rev. Mr. Image an examination of the original specimen figured by Parkinson; and in Mr. Stutchbury's collection there is a similar and more perfect specimen; both of these resemble *C. diadema* in general form, but have been too much worn to be positively identified. The following description is drawn up from some compartments collected by Mr. Searles Wood, belonging certainly to three and probably to four individuals, one of which was young; as these specimens agree in all essential respects, I feel pretty confident that the characters, by which the present species differ from *C. diadema*, are of specific value.

Structure of Shell.—The longitudinal ribs on each compartment (*i. e.* the circumferential transverse loops), are convex and prominent, as in *C. diadema*, but they are crossed by more prominent ridges of growth (fig. 8*a*, 8*e*) than even in the roughest varieties of that species, so that the surface of the shell is more rugged. In the three recent species—viz., *C. diadema*, *balænaris*, and *reginæ*, the surface of the wall all round the cavities occupied by the whale's skin, is striated only by very fine longitudinal lines; but here, the outer portion, or that (fig. 8*d*) formed by the transverse loops, is crossed by transverse ridges of growth, like, but less prominent, than those on the external surface of the shell. The minute teeth, along the lines of junction between the transverse loops, are here less regular, and can hardly be said to exist; for the two edges are locked together by what may be more strictly called minute zig-zag ridges (fig. 8*d*, 8*e*), than teeth. The exact number of the circumferential plications (fig. 8*b*) in the wall of the shell is variable, in the same manner as in the three recent species. In the rostrum which has been figured (8*b*, enlarged twice its natural size), there is a peculiarity, probably accidental, which I have seen in no other specimen—namely, that one of the transverse circumferential loops at the end of one of the original folds of the wall, has ceased to be added to, and therefore may be seen (rather on the right hand of the middle of the figure) to terminate in one of the cavities between two adjoining folds. The sutural edges of the compound radii (*d*, fig. 8*b*) are about as thick as, or rather thicker than, in *C. diadema;* for in the middle part they do not reach to the sheath by about half the thickness of the compartment. In the same manner as in *C. diadema* and *reginæ*, each ala here rests, not on the internal surface (as in *C. balænaris*, and in all other Balanidæ) of the radius, but on a special plate (*c*, fig. 8*b*, 8*c*); but in *C. barbara*, instead of there being a deep chamber, running up to the apex of the compartment, between the radius and the special plate, this part is filled up almost entirely by solid shell. Although the extent to which this chamber is filled up varies a little, and although its depth varies a little in *C. diadema*, yet there is a marked difference between the specimens of this latter species, in which the chamber is most filled up, and those of *C. barbara*, in which it is least filled up. The alæ are thick, as in *C. diadema*, and their sutural edges have a central ridge, sending off on both sides sinuous crests. The basal margins of the alæ are not short compared with their upper margins, and therefore the whole ala is not wedge-formed (fig. 8*c*); and in this rather important respect *C. barbara* resembles *C. balænaris*, and differs from *C. diadema*. The lower edge of the sheath does not seem to have projected freely,—in this respect, also, resembling *C. balænaris*. From the basal margin of the alæ not being narrow, and from the inner ends of the folded walls being, as it would appear, also broad, I have little doubt that the cavity in which the animal's body was lodged, resembled in shape that in *C. balænaris*, the membranous basis being larger than the orifice of the shell.

Opercular valves unknown.

Summary.—This species is more nearly related to *C. diadema* than to the others; but in some points, just specified, it resembles *C. balænaris*. The characters by which it differs from all the species are, firstly, the more prominent transverse ridges on the external surface of the shell, and more especially on the surfaces bounding the outer sides of the cavities occupied by the epidermis of the whale. Secondly, the character of the teeth, or rather ridges, along the lines of junction between the transverse loops. And, thirdly, the spaces between the radii and the special plates on which the alæ rest, being solidly filled up.

The *Coronula bifida* is an Italian tertiary species, so named by Bronn, in his "Italiens Tertiär-Gebilde" (1831), p. 126. It is very possible that this may be identical with *C. barbara*, but Bronn does not seem to have been aware of the absolute necessity of giving minute details in his descriptions of fossil cirripedes. The chief character of *C. bifida* is thus given :—"Eine tiefe Furcle oder Spalte theilt die Längenrippe von oben herab bis zur Halfte, welche bei der sonst ähnlichen *C. diadema* entweder ganz fehlt, oder nur zuweilen kurz angedeutet ist." Had it been stated that the longitudinal ribs were divided from the middle down to the base, instead of from the top to the middle, the description would have been intelligible to me, though the character thus afforded would not have been of specific value, as this dividing of the ribs occasionally occurs in all four species, and is produced by the formation of new folds in the walls.

Family—VERRUCIDÆ.

Cirripedia sine pedunculo : scuta et terga, musculis depressoribus non instructa, ex uno latere tantum mobilia, ex altero cum carinâ et rostro in testam asymmetricam immobiliter conjuncta.

Cirripedia without a peduncle : scuta and terga, not furnished with depressor muscles, moveable only on one side, on the other side united immoveably with the rostrum and carina into an unsymmetrical shell.

Genus—Verruca.

Verruca.[1] *Schumacher.* Essai d'un Nouveau Syst. Class., 1817.
Clysia. *Leach.* Journal de Physique, tom. 85, July, 1817; *Clisia,* Leach, Encyclop. Brit. Suppl., vol. iii, 1824; *Clitia,* G. B. Sowerby, Genera of Recent and Fossil Shells.
Creusia. *Lamarck.* Animaux sans Vertèbres, 1818.
Ochthosia. *Ranzani.* Memoire di Storia Nat., 1820.
Lepas et Balanus Auctorum.

The family of Verrucidæ includes only the above single genus; but it has, I think, as good a claim to be considered a distinct Family as either the Balanidæ or Lepadidæ, that is, either the Sessile or Pedunculated Cirripedes. The two latter Families differ from each other almost exclusively in the nature of the shell or external covering, and in the muscles moving the different portions of it: now Verruca has a very peculiar shell, destitute of all muscles, excepting the adductor scutorum, and composed of only six valves, and these are so unequally developed, that the longitudinal dorso-ventral plane of the body comes to lie nearly parallel to the surface of attachment, instead of at right angles to it. Upon the whole, the Verrucidæ are nearly equally related to the Lepadidæ and Balanidæ; but certainly nearer to the Lepadidæ, than to the sub-family Balaninæ or typical sessile cirripedes; though, on the other hand, if compelled to place Verruca in one of these two Families, I should place it amongst the Chthamalinæ, the other sub-family of Balanidæ. The distinctness of Verruca, though in appearance a sessile cirripede, from the Balanidæ or true sessile cirripedes, is interesting, inasmuch as no member of this latter Family has hitherto been found fossil in any Secondary Deposit, whereas Verruca ranges from the present day to the upper beds of the Chalk near Norwich, and in Belgium; being likewise found in the Glacial Deposits, in the Red and Coralline Crags of England, and in an ancient tertiary formation of Patagonia.

The shell of Verruca has generally been quite misunderstood: it consists, as already stated, of six valves; and these can be proved (as I have shown in my volume published by the Ray Society), by tracing the development of the young shell, to consist of a rostrum and carina, unequally developed on their two sides,—of a scutum and tergum in their normal and moveable condition,—and, lastly, of the scutum and tergum on the opposite side, most singularly modified, immoveably articulated to the rostrum and carina, forming together with them a shell, which is firmly united to the basal membrane, and so to the surface of attachment. It can be shown that the very remarkable modification and

[1] According to Bock, in the 'Naturforscher' of 1778, this term was used by Rumph for a Chelonobia, but as it was before the adoption of the binomial nomenclature, according to the Rules, it may be passed over, and does not interfere with the priority of Schumacher.

enlargement of the fixed scutum and tergum, is due to the development of a single small portion in each valve, namely, the lower ridge of the articular ridges by which these valves are united together. It is very remarkable that in all the species it seems to be a matter of chance, whether the right or left hand valves undergo this singular modification; consequently, of every valve it is equally likely to find a right-hand or left-hand specimen; and these, though exactly alike, except in being reversed, or in coming from opposite sides of the body, yet, from this very circumstance, and from the fixed valves being of very irregular shapes, are rather perplexing to identify. This short description will, I hope, suffice to make the following descriptions intelligible.

1. VERRUCA STRÖMIA. Tab. II, fig. 9 *a*, 9 *b*.

LEPAS STRÖMIA. *O. Müller*. Zoolog. Dan. Prod., No. 3025, 1776.
— — *Ib*. Zoolog. Dan., vol. iii, Tab. 94, 1789.
— STRIATA. *Pennant*. British Zoology, vol. iv, Tab. 38, fig. 7, 1777.
DIE WARZENFORMIGE MEEREICHEL. *Spengler*. Schriften der Berlin. Gesell., 1 B., Tab. 5, fig. 1—3, 1780.
LEPAS VERRUCA. *Spengler*. Skrifter af Naturhist. Selskabet, 1 B., 1790.
— — ET STRÖMIA. *Gmelin*. Syst. Nat., 1789.
BALANUS VERUCA. *Bruguière*. Encyclop. Meth., 1789; *Clisia verrucosa*, Deshayes, in Tab.
— INTERTEXTUS. *Pulteney*. Catalogue of Shells of Dorsetshire, 1799.
LEPAS STRIATUS. *Montagu*. Test. Brit., 1803.
— VERRUCA. *Wood's* General Conchology, Pl. 9, fig. 5, 1815.
VERRUCA STRÖMII. *Schumacher*. Essai d'un Nouveau Syst. Class., 1817.
CREUSIA STRÖMIA ET VERRUCA. *Lamarck*. Animaux sans Vertèbres, 1818.
OCHTHOSIA STROEMIA. *Ranzani*. Memoire di Storia Nat., 1820.
CLISIA STRIATA. *Leach*. Encyclop. Brit. Suppl., vol. iii (sine descript.), 1824.
CLITIA VERRUCA. *G. B. Sowerby*. Genera of Recent and Fossil Shells, Plate.
VERRUCA STRÖMII. *J. E. Gray*. Annals of Philosophy (new series), vol. x, Aug. 1825.

V. scuto mobili, cristâ articulari inferiore dimidiam brevis cristæ articularis superioris latitudinem non æquante: testâ plerumque longitudinaliter sulcatâ.

Moveable scutum, with the lower articular ridge not half as broad as the short upper articular ridge: shell generally ribbed longitudinally.

Fossil in Glacial deposits of Scotland, Mus. Lyell; Red Crag (Walton, Essex), Coralline Crag (Sutton), Mus. S. V. Wood.

Recent on the shores of Great Britain and Ireland; Shetland Islands; Denmark; Iceland; shores of northern Europe; Red Sea. Attached to shells, laminariæ, rocks, crabs, and floating bark, from low tidal mark to fifty or ninety fathoms.

I have seen a perfect specimen of this species from the Glacial deposits of Scotland, and separated valves from the Red and Coralline Crags, collected by Mr. S. Wood. The moveable opercular valves have not been discovered; and these are certainly much the most important parts for the diagnosis of the species; but the other valves are tolerably perfect, and are undistinguishable from recent specimens of *V. Strömia;* therefore, I have ventured, with some hesitation, thus to name these specimens. The fossil specimens all belong to the common variety, having its shell longitudinally ribbed, a character not observed in the four other species of the genus. As an aid to collectors in the Crag, I have thought it would be more serviceable to give a drawing (fig. 9 *a*), from a recent specimen, of all the six valves, separated, but in as nearly as possible their proper relative positions, and likewise of the under side of the fixed scutum and tergum, than to give fac-similes of such valves, in themselves not perfectly characteristic, which have as yet been discovered fossil.

It should be borne in mind, that of the six valves of which figures are here given, it is just as likely that reversed specimens from the opposite side of the body should be found, as these which represent valves taken from a specimen in which the left-hand scutum and tergum were fixed and formed part of the shell.

2. VERRUCA PRISCA. Tab. II, fig. 10*a*—10*c*.

VERRUCA PRISCA. *Bosquet.* Monographie des Crustacés fossiles du Terrain Crét. de Limbourg, Tab. 1, fig. 1—6; 1853.

V. testâ lævi; scuti mobilis cristâ articulari inferiore aliquanto latiore quam superior.

Shell smooth: moveable scutum, with the lower articular ridge somewhat broader than the upper articular ridge.

Fossil in Chalk, Norwich, Mus. J. de C. Sowerby: 'Système Senonien et Maestrichtien,' Belgium, Mus. Bosquet.

M. Bosquet has admirably figured and described the several separated valves belonging to this species, and I owe to his great kindness an examination of some of them. In Mr. J. de C. Sowerby's collection, also, there is a single specimen (fig, 10*a*), attached to a Mollusc, with the four valves of the shell united together, but without the two moveable opercular valves; it cannot be positively asserted that this is the same species with that of M. Bosquet, but such probably is the case. The opercular valves (fig. 10*b*, 10*c*) are necessarily figured from Belgian specimens. It is the English specimen to which I alluded in the Introduction to my 'Monograph on Fossil Lepadidæ.' This species of Verruca is interesting, from being the only known Secondary one, but in itself it is a very poorly characterised form, and I can point out no important character in the shell by which it can be recognised. The rostrum and carina, which are of nearly equal sizes, are locked together, and likewise to the fixed scutum and tergum, by the usual interfolding plates; the plates between these latter valves seem to have been less developed in M. Bosquet's specimen than in the English. The fixed scutum has a large adductor plate, which seems to have been chipped in M. Bosquet's specimen; this valve and the fixed tergum in all essential respects resemble the same valves in *V. Strömia.* The surface of the shell is very smooth.

The *moveable scutum* has its occludent margin considerably arched: the lower articular ridge is broader than the upper ridge, in which respect it resembles the same valve in *V. lævigata,* a South American form, but the whole valve is not so broad as in that species. There is no adductor ridge on the under surface. The *moveable* tergum has its upper articular ridge narrow, and slightly produced into a point on the scutal margin: in this latter respect this species, also, resembles *V. lævigata,* but differs from it in the whole valve not being so broad in proportion to its height.

TAB. I.

Fig. 1*a*, Balanus tintinnabulum, nat. size; small specimen.

1*b*, ,, ,, rostrum, internal view of, large specimen, nat. size.

1*c*, 1*d*, ,, ,, scutum and tergum, internal views of, from recent specimens, the opercular valves not having been found fossil.

Fig. 2*a*, Balanus calceolus, engraved from a recent specimen, the single fossil having been injured during examination.

2*b*, 2*c*, ,, ,, scutum and tergum, internal views.

2*d*, Spur of tergum, variety: all from recent specimens, the opercular valves not having been found fossil.

Fig. 3*a*, 3*b*, Balanus spongicola, scutum and tergum, external views.

3*c*, 3*d*, ,, ,, ,, ,, internal views.

3*e*, ,, ,, shell, enlarged from a recent specimen, the single fossil being young, and having been injured by examination.

Fig. 4*a*, Balanus concavus, shell (Coralline Crag specimen).

4*b*, ,, ,, internal view of part of the carina (to the left hand), of the carino-lateral compartment, and of part of the lateral compartment, showing the raised borders (*r*) on the rostral sides of the sutures in the sheath (Italian Tertiary specimen).

4*c*, ,, ,, shell, *var.*, with longitudinal ridges or ribs not prominent (Coralline Crag).

4*d*, ,, ,, smooth *var.* (Piedmont specimen).

4*e*, ,, ,, scutum, external view (Coralline Crag).

4*f*, ,, ,, ,, internal view do.

4*g*, ,, ,, tergum, external view do.

4*h*, ,, ,, scutum, internal view (Maryland, U.S.).

4*i*, ,, ,, tergum, internal view do.

4*k*, ,, ,, ,, external view do.

4 *l*, ,, ,, scutum, external view (Portugal fossil, and recent).

4*m*, ,, ,, tergum, external view do. do

4*n*, ,, ,, scutum, internal view do. do.

4*o*, ,, ,, tergum, external view, very large specimen (Turin).

4*p*, ,, ,, scutum, small portion, at the rostral corner, highly magnified, very large specimen (Turin).

Fig. 5*a*, Balanus porcatus, shell, nat. size (Red Crag).

5*b*, Balanus porcatus, small portion of basal margin of wall, much enlarged.
5*c*, ,, ,, portion of basal plate, much enlarged.
5*d*, 5*e*, 5*f*, 5*g*, Balanus porcatus, scutum and tergum, external and internal views; engraved from recent specimens, the opercular valves not having been positively found fossil.

Fig. 6*a*, Balanus crenatus, twice nat. size (Red Crag).
6*b*, ,, ,, small portion of basal margin of wall, much enlarged.
6*c*, ,, ,, portion of basal plate, of unusual thickness, much enlarged.
6*d*, 6*e*, ,, ,, scutum and tergum, external views from recent specimens.
6*f*, 6*g*, ,, ,, ,, ,, *var.*, internal views from recent specimens.

Fig. 7*a*, 7*b*, Balanus Hameri, scutum and tergum, external views engraved from Canada glacial specimens, the opercular valves not having been found in Great Britain.
7*c*, 7*d*, ,, ,, scutum and tergum, internal views of very large and old specimens (Canada and Uddevalla).

Tab I

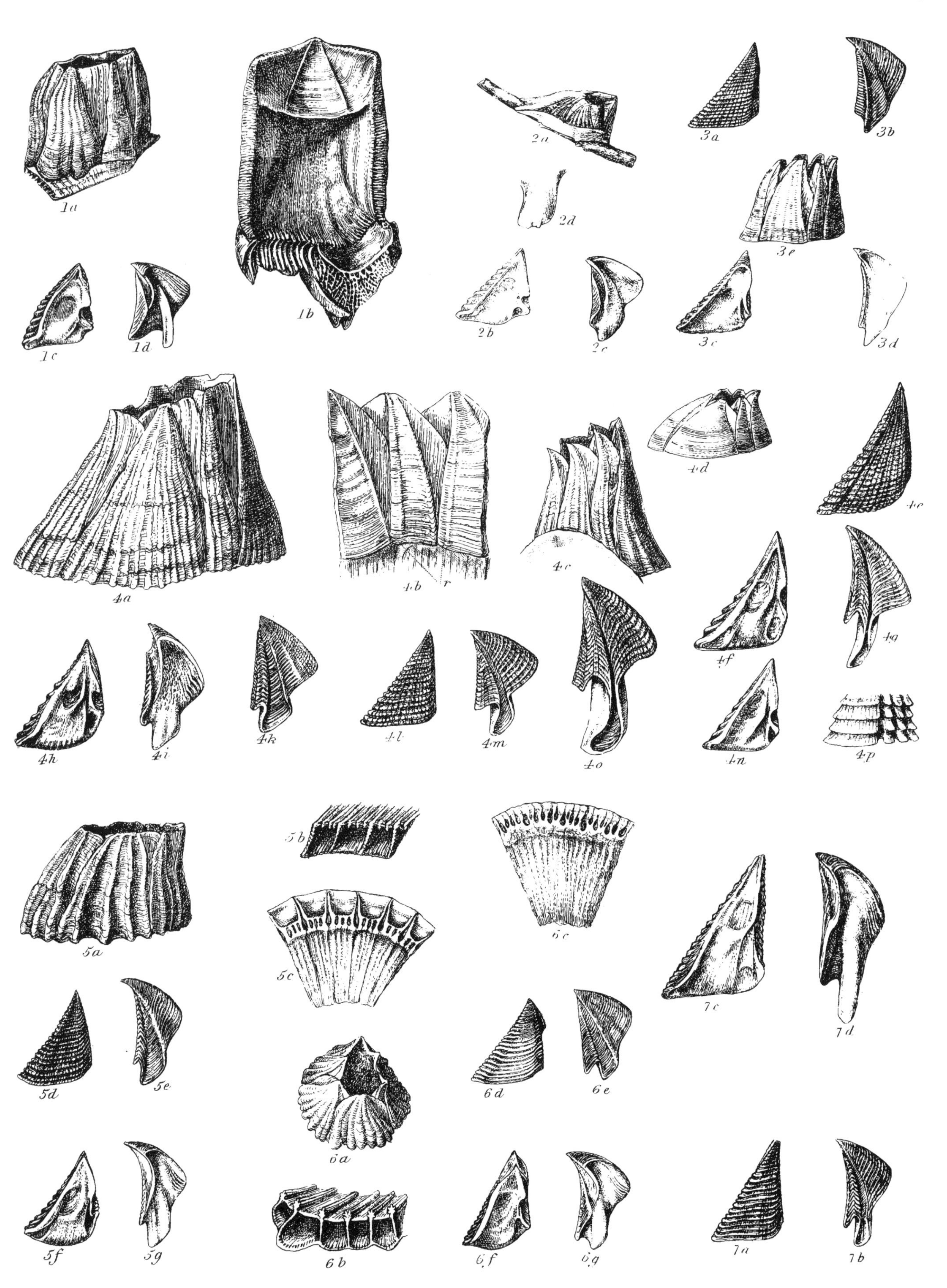

B A L A N U S.

TAB. II.

Fig. 1*a*, Balanus Hameri, lateral compartment, viewed externally.
1*b*, ,, ,, carino-lateral compartment, viewed internally.

Fig. 2*a*, Balanus bisulcatus, shell, $1\frac{1}{2}$ nat. size.
2*b*, ,, ,, rostrum, viewed externally, to show the commencement of the folding of the walls, $1\frac{1}{2}$ nat. size.
2*c*, ,, ,, *var.* plicatus, thrice nat. size.
2*d*, ,, ,, very young shell, probably belonging to this species, enlarged about seven times.
2*e*, 2*f*, ,, ,, scutum and tergum, external views.
2*g*, 2*h* ,, ,, ,, ,, internal views.

Fig. 3*a*, Balanus dolosus, rostrum, much enlarged, viewed internally.
3*b*, ,, ,, portion of basal plate, much enlarged.
3*c*, 3*d*, ,, ,, scutum and tergum, external views.

Fig. 4*a*, Balanus unguiformis, shell, twice nat. size.
4*b*, ,, ,, shell, *var.* erisma, twice nat. size.
4*c*, 4*d*, ,, ,, scutum and tergum, external views.
4*e*, 4*f*, ,, ,, ,, ,, internal views.

Fig. 5*a*, Balanus inclusus, shell, nearly thrice nat. size.
5*b*, ,, ,, basis of ditto, showing basal edges of the compartments.
5*c*, ,, ,, *var.*, with its rostro-carinal axis elongated.
5*d*, ,, ,, *var.*, ,, ,, ,, showing the narrow furrowed basis and lower portions of the six compartments.
5*e*, ,, ,, *var.* intermediate between the last two varieties, basal view of, nearly twice nat. size.
5*f*, 5*g*, ,, ,, scutum and tergum, internal views from a very young specimen.

Fig. 6*a*, Acasta undulata, rostrum viewed externally; spinose *var.*; twice nat. size.
6*b*, ,, ,, lateral compartment, smooth *var.*, twice nat. size.
6*c*, ,, ,, basal cup.
6*d*, ,, ,, scutum viewed internally.
6*e*, 6*f*,, ,, ,, and tergum viewed externally.

Fig. 7*a*, Pyrgoma Anglicum, shell, viewed from above, four times nat. size.
7*b*, ,, ,, scutum, internal view.
7*c*, ,, ,, tergum, internal view, engraved from recent specimen, this valve not having been found fossil.

Fig. 8*a*, Coronula barbara, rostrum, external view; nat. size.
8*b*, ,, ,, ,, viewed internally; twice nat. size.
8*c*, ,, ,, lateral compartment, internal view.
8*d*, ,, ,, internal view, much enlarged, of small portion of basal margin of folded wall.
8*e*, ,, ,, external view, greatly enlarged, of small portion of surface of folded walls, near the basal margin; (*r r*) the transverse ridges of growth.

The following letters of reference apply to all the figures of Coronula:

a, sheath marked transversely in the upper part by the attachment of the opercular membrane.
a', ala
b, furrow on each side of (*a*), receiving the edge of the thick ala of the adjoining lateral compartment.
c, special plate, on which the ala rests.
d, radius, on the edge it may be just seen to consist of an outer layer (the normal radius), and a much thicker inner part (the pseudo or complementary radius) formed of oblique denticulated septa.
e e' e'', basal edge of wall, which from its commencement at *e*, or *e''*, can be followed, folding up to near the basal edge of the sheath, to its termination at *e''* or *e*.
f f, serrated lines of junction between the folds of the wall.

Fig. 9*a*, Verruca Strömia, much enlarged, engraved from a recent specimen, only certain valves having been found fossil.
9*b*, ,, ,, fixed scutum and tergum, internal views. The following letters apply to both these figures:

A, rostrum.
B, carina.
S, moveable scutum, S', scutum fixed and modified so as to form part of shell.
T, moveable tergum, T', tergum fixed, forming part of shell.
In S, and S', *a* is the occludent margin; *b*, the basal margin; *m*, the plate to which the adductor muscle is fixed.
In S and S', the tergal margin is marked by small dashes; (') being the upper articular ridge, and ('') the second or lower articular ridge: in S' ('') is called the parietal portion of the valve.
In T and T', the scutal margin is marked by small dashes; (') being the first and upper articular ridge, hardly distinct from the occludent margin, and called in T' the occludent rim; ('') is the second or middle, and (''') the lower or third articular ridge, called in T' the parietal portion of the valve: *x* is the carinal margin, called in T' the carinal rim, and the basal margin.

Fig. 10*a*, Verruca prisca, five times nat. size.
10*b*, 10*c*, ,, scutum and tergum, external views; engraved from a Belgian Cretacean specimen, the opercular valves not having been found in England.

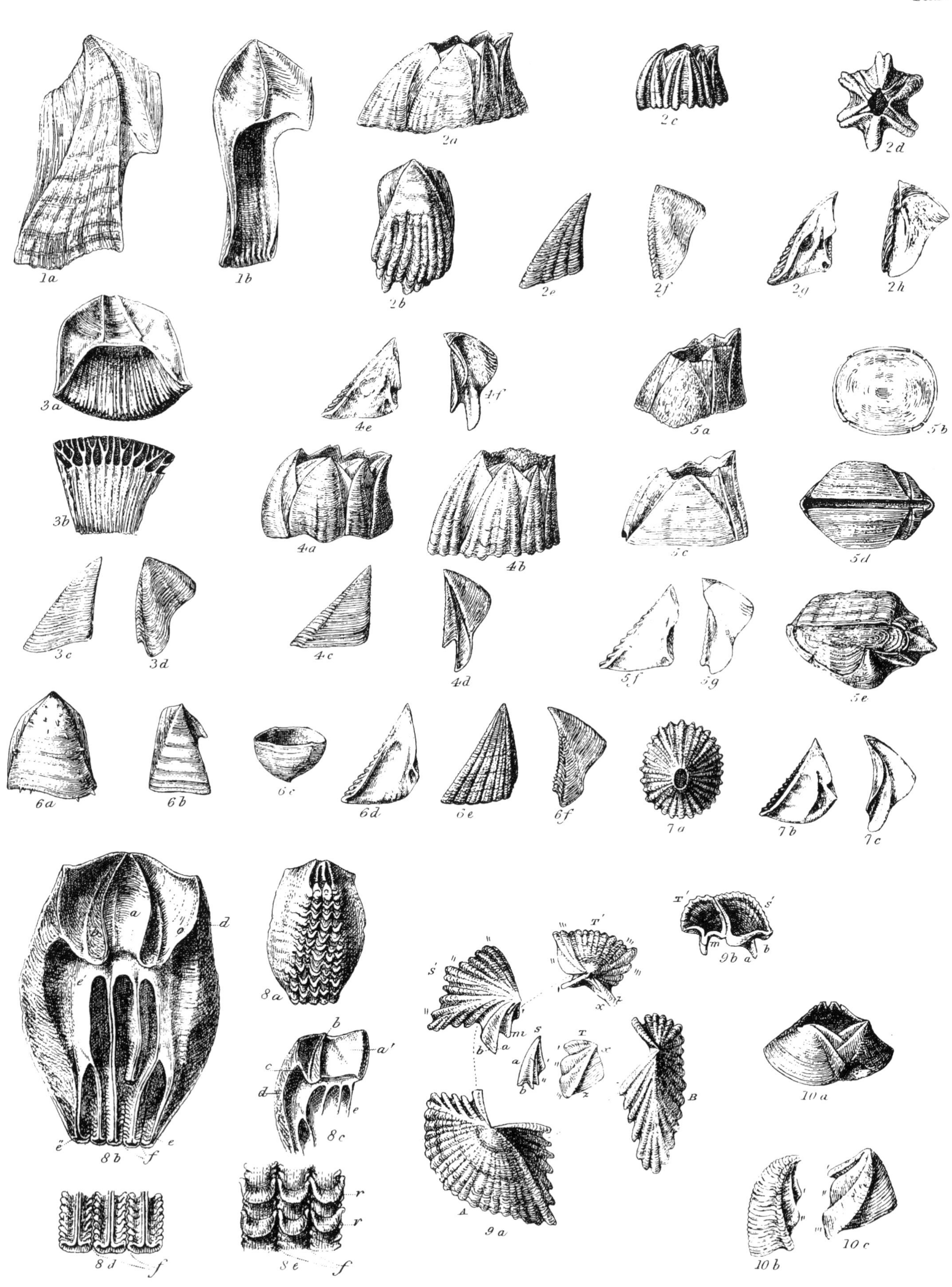

BALANUS, ACASTA, PYRGOMA, CORONULA, VERRUCA.

INDEX

TO

MONOGRAPH ON FOSSIL BALANIDÆ.

BY

C. DARWIN, M.A., F.R.S., &c.

N.B.—*The names in Italics are Synonyms.*